Artificial Intelligence

A Practical Guide to Improving
Your Life With Ai

*(Artificial Intelligence for Business and Computer
Networking for Beginners)*

Emily Cantrell

Published By **Simon Dough**

Emily Cantrell

Artificial Intelligence: A Practical Guide to Improving Your Life With Ai (Artificial Intelligence for Business and Computer Networking for Beginners)

ISBN   978-1-998901-49-4

No part of this guidebook shall be reproduced in any form without permission in writing from the publisher except in the case of brief quotations embodied in critical articles or reviews.

Legal & Disclaimer

Table Of Contents

Chapter 1: What is Artificial Intelligence?

There are numerous types of technology available there for businesses to utilize. They may use programming to aid in the development of programs of their own and ensure that they're set up to meet the diverse needs of their clients. They can use mobile phones and tablets and databases to ensure they get the results they desire and more. One aspect that is really taking off in terms of how companies can complete their job and provide excellent service to their clients is by using artificial intelligence.

When people hear the term artificial intelligence (AI), their minds immediately goes to computers that could rule the world and take over in ways normally reserved for human beings. Although it is true that various sub-sects of artificial intelligence can teach computers and other systems to think that is similar to human brains but artificial intelligence isn't necessarily going to compare with the kind of technology we've seen in sci-fi flicks we may have enjoyed during the previous time.

This being said, it's crucial to gain a solid understanding of the way artificial intelligence operates as well as how to apply it to benefit any type of business and other fields that we might want. Many different components comprise artificial intelligence, and understanding the whole process and how it functions, could be the first step towards using it.

For a start the discussion, artificial intelligence, which is commonly known as AI is a field in computer science. This can be defined as the creation of intelligent machines and systems that work and, often, react like what we experience in humans. Many computer tasks rely on AI are designed to aid out in areas such as the planning, learning, problem solving and speech recognition.

In essence, when working on artificial intelligence, we will work on a specific part of the wider area in computer science. In this specific area we will be investigating a method that could create intelligent machines that can think and behave independently. Due to the capabilities of these machines, they have rapidly become an integral component of the technology industry , as increasing numbers of firms and businesses utilize this in their daily operations.

The research conducted on AI is highly technical. This is due to the way artificial intelligence functions as well as the various components that it comes with. The most fundamental issues that arise from the AI are the possibility of creating programs to program the computer for some specific trait. This includes the ability to manipulate and move a range of objects, learn the ability to perceive, plan and knowledge, problem-solving and reasoning.

The power of knowledge will be a factor when dealing with AI, and knowledge engineering is likely to form the foundation of the research field. Machines can often learn to behave and behave as a human can, but only when they have received enough details about the way that the world functions and how they ought to behave.

Keep in mind that with artificial intelligence, your computer isn't simply sitting there learning without assistance. The computer will be limited by the data and examples you provide to it, and the amount of learning it can do in this period. If incorrect answers or examples are provided to the computer system or it is not properly trained and in the correct way, it can become very problematic. The machine needs precise information, and plenty of it, to act and behave in a way like what we experience in humans.

Artificial intelligence, to ensure that it is effective, must be able to access things such as relationships and properties, categories and objects that are a part of all them in order to realize the concept of knowledge. This is the reason this kind of thing will not be carried out all the time. The ability to instruct computers how they are supposed to respond to everything around it and to use thinking, problem solving or common sense a job that is extremely difficult and laborious. Although it's amazing the capabilities this technology can accomplish, AI does not make sense in terms of cost and time required for each project you'd like to work with.

Machine learning is a different well-known technology is being explored by a lot of developers and is the core element of the process that is used to create artificial intelligence. Learning with no supervision will have the robot develop the ability to spot patterns in data streams that it sees.All that learning, when it is under the proper supervision will involve both numerical regressions and classifications.

What does this all have to do with it? The first is that the classification issues could be useful since they can be used to determine the place an object falls in to, and which category is most crucial for it. In addition, there are regression problems which are designed to require an array of input and output instances that have a numerical. This allows us to find the appropriate functions to facilitate the generation of the right outputs using the inputs we are in a position to locate.

There are plenty of things you're capable of doing with regards to artificial intelligence and machine learning but this isn't the only area that is associated with AI. You could realize that robotics is likely to be a significant subject that we must look at in addition. Robots will require an intelligence level to manage many of the tasks expected of robots, such as the ability to manipulate objects as well as perform any type of navigation.

To simplify this process Artificial intelligence is expected help the machines you run and programs to gain knowledge from their own experiences. This aids these machines in being capable of adapting to any new inputs it receives which results in the execution of tasks that resemble human. The majority of the cases we hear about regarding AI currently use a combination of natural machine learning and language processing. When we employ these techniques in our jobs it is possible for the computer to be taught to perform the job in the best way. It is simply by learning the proper algorithm to process an enormous volume of data, and also being able to recognize patterns in the data.

Let's pause here for a minute and go through some of the background associated with artificial intelligence. The term "AI" was first spotted in 1956, however in the past decade or then, AI has become more widely used. The rise of popularity can be attributed to advancements made by the power and storage of computers as well as some of the most sophisticated algorithms that are out are available, as well as the increase in volume of data.

At the beginning of the research conducted on AI the subjects studied in this regard included the various ways used to represent as well as some methods of problem-solving. In the 1960s in the 1960s, US Department of Defense started to become interested in this field and the things it could accomplish and this led to their use of AI to instruct their machines and computers to imitate some of the fundamental reasoning abilities we observe in human beings.

A portion of this research was able to open the way to automate formal reasoning as we can see in machines we have in the present. These include smart search systems and decision support systems, which can be developed to enhance and improve the capabilities are present in machines that are modeled after what can be done by humans.

There are many science fiction or films in Hollywood and see how AI is presented in a manner that suggests robots which are similar to humans are dominating the world. You can be sure that the advancements and developments taking place with AI will not be as frightening or even as intelligent. AI is an intelligent instrument, but it cannot be used to allow computers and machines conquer the world. However, AI has evolved to bring a myriad of benefits for almost every field that exists. We'll look at these further in this guidebook however for now, many industries such as marketing, retail and health care are utilizing the elements of artificial intelligence to accomplish their tasks more effectively and efficiently more effectively than they did before.

The second thing we should look at is the reason why artificial intelligence is important. Why should we invest so many hours looking through the basic concepts of artificial intelligence? And why this topic is important at all? Artificial intelligence is beginning to dominate the world and understanding how to apply it to purposes can make a significant impact on how companies serve their clients, and many more. There are numerous reasons why artificial intelligence will be a crucial thing to master. The reasons include:

AI can automate the process of repetitive learning processes that take place and could make more discoveries from data. AI will differ from the methods we have seen with robotic automation that is driven by hardware. Instead of going through and automate some tasks that require manual effort, AI is going to perform frequent, computerized and large-scale tasks efficiently and without fatigue. In order to get this type of automation in place however, human interaction is crucial in setting up the system correctly and asking the right questions which are required.

Additionally, AI is going to ensure that a product that you already have will have additional technology added. Most of the time, AI is not going to be sold through the aid of an individual application. Instead, the software you currently use can be enhanced through the use of AI. Numerous intelligent machines are able to be combined with the information that companies hold onto in order to ensure that technology is upgraded regardless of what sector it's in.

AI will be able of adapting to the changing algorithms of learning to enable the data to be able to program. AI can detect any irregularities or patterns in the data to ensure that the algorithm can acquire the ability to learn. This algorithm can transform into predictive or a classifier. In the same way that we observe the algorithm being capable of teaching itself how to play chess, it will also be equipped to utilize this kind of concept to suggest which product to purchase in the future.

The models will be able of adapting as they receive new information. Backpropagation will be among the methods used by AI employs to ensure that the model chosen can be adjusted with training and additional information if the first answer given isn't quite right.

AI can then examine more detailed data and more of it thanks to the aid of neural networks which have many hidden layers. For instance, using the ability to identify fraud was a challenge in the past, but now, thanks to powerful computers and AI as well as massive data, it's something that banking and finance institutions have been now able to focus on. Of course, to be certain that this type of model is effective is to have a huge amount of data that it can be trained on. In reality the more data can be fed into the model, the more precise the models will become.

The next thing to consider is the fact that AI will be extremely accurate in what it can do with the aid of deep neural networks. This was practically impossible prior to. For instance, all of the interactions you make using Alexa or other similar applications, are likely to be based on capabilities that can be achieved by using deep learning. As we continue to make use of them, the more precision will be evident in them, too.

As compared to other ways for analyzing and comprehending information, AI can be the one that is capable of getting the most from the data. If the algorithms can learn by themselves then the data will be transformed into intellectual property. The answers the system requires are likely to be located in the data, and using AI can aid in the discovery of these. Because the importance of data is viewed as much more significant than before, this will assist a business to learn how to gain an advantage over other companies. If you're able to obtain the top data available in an industry which is filled with methods that are similar, the most accurate data is helping you to be the top performer.

We have a better understanding benefits of AI that can be derived from a myriad of needs that can handle every piece of data you're using, it's the time to look at the ways you can make use of AI and also various sectors that can benefit from the various capabilities of AI. Although it might seem that AI could be something that should be restricted to the technology and science fields however, it is actually possible to make use of the power of AI to serve every business that exists. This is especially true when we are in requirement for a system to answer questions which can be utilized to meet a range of requirements such as medical research, risk notifications patent searches, as well as legal assistance.

Let's look at the various industries that can be worked with AI as well as all of the possibilities that are available with this particular area of computing science. The first one is healthcare. The various applications that use AI will provide patients with customized X-ray readings and also medicine. Personal health assistants could be used as life coaches to encourage us to eat better and exercise regularly, as well as taking medication at the appropriate moment with AI capabilities.

Retailers also stand to gain the use of AI. AI will really come into play when it's time to offer a shopping experience that is digital, while providing recommendations which are personalised to the user. This could be used to talk about various options for purchasing that are offered to the customer. Site layout and management of stock technologies will improve with the many capabilities of AI.

The next item on the list will be manufacturing. It is possible for a variety of manufacturing firms to look through and employ AI to analyse the manufacturing IoT data that is streaming from the devices that are connected. This can be carried out to anticipate the demand and lead with the aid of what's called a recurrent system. Recurrent networks are likely to be a form of deep learning network which can be utilized with any type of sequence data . They can be extremely useful in this type of processing.

Finally, it is possible to use AI in order to assist in the banking sector. AI can increase the speed, accuracy and even the efficiency that human effort. For financial institutions they will be employed to assist us determine what transactions are most likely to be legitimate and which are not as well as how to analyze and apply fast and accurate credit scoring, as well as how to automate certain operations that require data management which are more demanding.

It is important to comprehend this process: we must to collaborate with AI. Artificial intelligence isn't an invention that's here to take over our jobs. It is instead coming to improve our capabilities and ensure that we're better at what we do. We're supposed to work in tandem with this AI instead of believing that it will be able to cause trouble and eventually replace us.

Since the algorithms available for AI will learn differently that humans do and also, they are will be able to see things and the world around them in a different way. This can be extremely useful because it permits the machine to spot patterns and connections in the vast amount of data which we may not notice. This AI-human collaboration is likely to provide numerous opportunities for the future, such as:

1. Apply analytics to some sectors and domains that aren't being fully utilized.

2. It is possible to enhance some of the efficiency we observe using the existing analytical technologies which include computer vision as well as the analysis of time series.

3. Help us break down the barriers to economic growth that exist present, and include obstacles that are created by the translation process and with language.

4. It's going to enhance some of the capabilities that exist and helps us do better in the work we're currently doing.

5. We need a clearer vision and understanding, better memory , and much more at the same time.

Although there are plenty of advantages to taking advantage of all the benefits artificial intelligence has to offer humans, there's going to be challenges which it will bring also. Artificial intelligence will provide the potential to shake all industries if it is utilized correctly. A lot of industries have begun to utilize the technology made available by AI and it's likely that more more industries will utilize these tools to assist them in getting more efficient in their work.

In this regard It is essential to every industry and business that is considering artificial intelligence to realize that the AI system has certain limitations. One of the major drawbacks that could be discovered when using AI is the fact that it will learn from data. It is the only method to incorporate the knowledge and data that we would like for the AI system to gather.

This means that if the data is comprised of areas that aren't completed or are not accurate, it's likely to be reflected in the results you receive. Any extra layers of analysis and predictions must be added to this in a separate method.

Nowadays, the systems that are deemed to be AI will be taught in a way which assists them in performing an activity that is specifically identified. The AI system which can be taught playing poker by itself cannot use this knowledge to play solitaire or chess. It is possible to learn to teach it these games if we'd prefer, but it's not like us and cannot use the knowledge it acquired while learning to play poker to play other games. You must undergo the process of learning over and over and learn it in a variety of way, and with ample room in the memory in the machine, to allow the machine or computer to master multiple games.

Additionally the system that was designed to aid banks and other financial institutions spot any fraudulent activity being committed cannot be used to drive a vehicle or give good legal guidance. However, even within the same field of job the AI system will only be able of doing only one thing. For instance, you could be using an AI capable of identify fraud in the health sector, but it isn't able to find out if there is fraud on tax or warranty claims fraud.

Also, they will be systems that must be extremely specialized to function. They will focus on only one task, and are not likely to be capable of learning and adapting and behave in a way that is identical to humans. It will be able to understand how to perform things, but generally it is a system that is specific to which means it can only learn to function with a specific type of system and that's all it can accomplish.

Even with these limitations however, there are plenty of other things you can accomplish using AI and these could help you in bringing your AI system to perform the tasks your company requires. What exactly is artificial intelligence operate in the first in the first place? AI can work by combining large amounts of data by using algorithms that are speedy and can perform iterative processing. It is implemented in a way that allows the program to learn in an automated way when it is learning from the characteristics or patterns that will appear in the data employed.

AI will be an expansive subject which will cover a wide range of industries, and much more. The field of study will comprise a wide range of techniques, methodologies and theories can be utilized. Alongside this, you'll be able to utilize one or more of the major subfields listed below to achieve the most value from artificial intelligence.

1. Machine learning is likely to be an artificial intelligence subset and will help make it easier to build analytical models. It will employ many different techniques to create these models which include operations research, physics statistical analysis, and neural networks. All of these are employed to discover any hidden insights within the data

without being programmed to figure out where to look or which it will be able to predict when making predictions.

2. A neural network. This will be an example of machine learning which has several interconnected units which are like what is found in the brain as well as neurons that process information by responding to the inputs provided. The information is then transmitted across the various units. This process will need us to make numerous times to look at the data in order to make sure that we're capable of identifying the connections . The system will determine a interpretation and predictions from data, which is not defined.

3. Deep learning. It will make use of an enormous amount of neural networks that comprise a variety of processing units. this lets it take advantage of advancements in computer technology and more effective training methods. All this is done to allow the machine to be able to recognize some of the

more complicated patterns that can be discovered in large amounts of data. Image recognition and speech recognition are just two applications that will make use of this.

4. Cognitive computing is a different sub-field of AI which will aid us in getting an organic interaction with machines. Utilizing AI that includes cognitive computing will have the aim of allowing the machine to simulate human behavior through the ability to read images and speech.

5. Computer Vision: This relies on pattern recognition, in conjunction with some deep learning to detect the images or videos that are displayed in a video or image. If machines can analyse, process and understand the image that they are viewing, they stand an increased chance of recording videos and images in real-time. Then, they can understand what is happening in the surrounding.

Alongside these in the future, there will be several other technologies that will help to not only allow AI but also to help aid in the implementation of it. We can first consider GPUs or graphic processing units. They will be an important part of the capabilities of AI as they are expected provide some of the larger computing power needed to aid in the iterative process. Learning neural networks are likely require lots of data that is big, with good computing capabilities to get the job accomplished.

The next aspect that AI requires support for is what is known by"the Internet of Things. This will allow for the generation of huge amounts of data generated by connected devices, and the majority of it isn't going to be analysed. It is possible to utilize the data more when we choose to automatize some of the models we employ using artificial intelligence, as well.

There will be certain AI algorithms which are sophisticated and are also more advanced. These are being developed and then merged with new methods to assist with the AI capabilities that you're seeking. All of this is as it allows us to process more data more quickly and can perform this at different levels. This kind of intelligent processing is going be the most important factor in being able to predict and identify unusual events that could occur as well as understanding the intricate systems that come the way and optimize situations that are unique at once.

Finally, APIs, also known as the interfaces for processing applications are going to be vital when it comes to implementing AI. They'll be a form of package of code that can be used in a variety of ways and will allow us to enhance the functionality of AI to the existing software programs and products. If you have existing software and other products that are available and being used, but you want to add some artificial intelligence without the need to begin from scratch This is the issue you should be focusing on.

These APIs will enable you to add to the diverse functions which are offered by AI that includes features for image recognition. Like what is required by security systems and Q&A capabilities which will assist in explaining the data, generate headlines and captions, and provide some intriguing patterns and insights that can be found within the information, however you may not have been able to capitalize on previously.

You can observe you will find a variety of diverse components with AI and many various things you're in a position to accomplish when working with this area that is computer science. Understanding how to use artificial intelligence and the many other features that it offers can allow you to improve your performance at work and help you get your work done efficiently and quickly and will help bring businesses forward by giving them an advantage over the competition.

Chapter 2: Making It Big with Big Data

The next subject we should take the time to study is the concept of "big data. It is the term used to refer to the huge amount of data, that can be either structured or unstructured as well, which is likely be available to businesses on a regular basis. While the majority of the time , we're talking about an enormous quantity of data, it isn't the main thing to consider when discussing this. What is most crucial here is how the company will be in a position to access this data and utilize it. The big data can be examined to help the business discover valuable insights that can help them make the most effective business decisions and even better decisions than they did before.

In the case of big data there are a variety of parts that work together to create the term. Although this concept, particularly, is relatively new, the idea of being able to collect and save a large amount of data to use to analyze data has been in use for a long time. Businesses have always been in the business of acquiring all the information they can and then use it to aid them in their. If they can analyze it in a proper method, and get the relevant insights, will help them gain an edge across the globe and aid them to be ahead in providing their customers with the best service.

In this regard, we can find couple of definitions that can be found to help us understand the nature of big data concerning and exactly how it functions. There will be three primary Vs can be used to assist us in getting more understanding of huge data, and how to make use of it for our own needs.

First, we have volume. This one is important because companies will invest some time in collecting their data, usually from a variety of sources. Making a decision based on data from only one source may appear as though it's simpler but it's not the best approach to making solid business decision-making. There are numerous ways in which information can be collected, such as from the business's transactions and social media, surveys and much more. In the past , the ability to keep this information in a safe place or even get access to it, could have been an issue. However, today there are many different platforms and technologies which can manage this information and allow companies to collect as much data as they want, without having to think about the location of the data.

The other V to pay attention to is speed. The information a business will collect can be streamed through at a high speed, something hasn't been experienced in the past and we must ensure that this data is dealt with now, not some time in the near future, but more timely way. Sensors and RFID tags are set assist us in getting the information we need and help it interact with us, working in close to real-time.

The third V is the word "variation. The most valuable kind of information is one that contains lots of diversity. This will guarantee that the company is receiving the information it requires and not getting caught in one direction in the process. The information that the business utilizes can be found in many different formats. It could be financial transactions, information from stock tickers and audio or video email messages, text documents which aren't structured more structured data, or certain numeric data we can find in a few traditional databases.

Although these three elements could provide us with some information about the big data that we are working with and the information we will be able to extract from this data set, there's two additional dimensions might be worth looking at in order to understand what big data is about and how it's going to be used. The two other dimensions will include the complexity and the variety of the data we're working with.

The primary dimension we'll examine is the variable. Apart from being able to boost both the types and velocity of data as well, information flow may often be extremely inconsistent and will be accompanied by periodic peaks that could cause problems with the data. This is often the case when discussing social media.

Although these peaks can be thought of as a positive factor because they mean that our clients are in the market and giving us more data that will help us improve however, these peaks, regardless of whether they occur during the season or daily, or even due to an event, can be difficult for companies to master. When these peaks are triggered by information that isn't structured and not structured, the problem is likely to get more severe.

The final aspect we'll examine in this section will be level of complexity. It is likely that the data that a company collects from its clients, services and products are likely be coming from multiple sources. This can make it challenging to cleanse transform, match, and even connect data from different systems. It is vital for a company to be capable of connecting and coordinating the hierarchies, relationships and data linkages in multiple ways otherwise your data will be spiraling into chaos in no time.

After we have some more of how this can be done now is the time to take a look at some of the reasons big data is important. What is the reason we are dedicating to a whole chapter in this book to the concept of big data? And how to make use of it. Why would other businesses need to have all this information and make use of the data for their own purposes?

The most important thing we're likely to discover when it comes to large datasets is that the importance does not depend on the amount of data available. Although it's true that the majority of companies will have tons of data, there are some with a bit less may gain an advantage due to their ability to make the most of the data accessible to them. What's most important when it comes to the data you have is what you choose on doing with your information that you are presented with.

It is feasible for businesses to bring into account data from any type of source and take the time to look over it to determine the information which are required. The answers you're searching for will depend on what type of company you operate and what your objectives and goals are. When this data is used in the right method, will assist us in reducing costs, decrease the time needed to finish the task and can assist us in optimizing the offerings and products that are available to clients, and guarantee that all decisions that are taken are wise choices.

If you can collect all the information that a business is collecting over a long period of time, and put it with a higher-powered analytics, there's going to be several various tasks associated with an organization that you're capable of accomplishing. The tasks that can be accomplished when you combine data and the analytics are:

Large data could be utilized to determine if there is a certain conduct that's considered to be fraud before it can be used to impact your company and create chaos.

Big data can be accessed and recalculate the risk in your portfolio in only one or two minutes.

Big data can be used to make shopping of your consumers more enjoyable. It can help in generating coupon codes at points of sales based upon the shopping habits observed by your clients.

Large data sets can be utilized to determine the causes behind issues, defects, or issues that occur in near-real-time throughout the process.

There are numerous companies and industries that don't have to worry about using large data in any way. They might be collecting data and analyzing it occasionally but they're worried that they're not able to utilize the data in a way that is appropriate or that it could require too much time. The companies are not utilizing their opportunities in many instances as the data remains in the data center without being utilized in the way it ought to be.

In reality, every industry will make use of this data to get the results they require to be competitive and stay ahead in the marketplace. Big data is likely to assist companies in virtually every sector that exists. Let's look at some industries that will benefit from this massive data, and also the methods they are actually using the data.

The first sector that has seen gain from the use of big data will be the banking industry. Consider all the information and data an institution or even a local one, has to have to deal with all the time. This is a huge quantity of data that is brought in from diverse sources. In light of this, banks are searching for new and creative ways to deal with their customers.

They also face the additional challenge of trying to ensure that their satisfaction with their customers is high , while decreasing their own risk and stopping fraud simultaneously. When you add this to the mix the fact that there are usually going to be laws on both a national and state level, making it difficult for banks to handle everything. With the aid of the massive data banks have, they are able to assist in meeting the various demands within one.

Another area where big data is expected to be crucial is in education. Teachers will often require an abundance of information in order to have an impact on not just the school system, but also on the course they select and the students. If the teachers, whether they are principals, teachers, or anyone other in the school, decide to study all the data that they can get and analyze it, they are able to accomplish their task.

Based on how the teacher decides to utilize the massive data and how they use it, they'll be more prepared to recognize students at risk and might require additional assistance. They can ensure that all students will be able to make the appropriate kinds of progress, and they could even take the necessary steps to ensure that there is the best system of assistance and evaluation of every teacher within schools.

The government sector is yet another one which could benefit from big data to improve the quality of life. Consider the huge quantities of information that governments can collect from citizens regarding the things they do and how they use it frequently. Our utility companies too will be able to use this type of information in order to figure out how much to be charged, where to put new utility services, and so on.

Governments can make use of a variety of analytical elements to help out with large information. When they're able to perform this in a proper way, we can observe that they are able to gain a amount of ground in preventing possibility of crime, dealing with congestion in traffic, overseeing all the agencies that are that are part of the government, or managing our utilities.

The next area that is most likely to benefit from our efforts to make use of big data will be the health care. Consider how much data is available on every particular patient within the health field but not the amount of information available when we put all the patients. Between records of prescriptions, medical patient records, or treatment programs, there's plenty of data that needs to be handled in this area.

In the health care industry there are things that must be completed as quickly and as accurately as is possible. and with sufficient transparency, as well as privacy of the patient and patient privacy, in order to comply with all the regulations within the field that are fairly stringent. If big data are used efficiently in this field, the service providers will be able to learn more about the patient's needs than they did before, and could be able uncover certain insights that remain hidden.

The manufacturing industry will not only benefit from the use of artificial intelligence as we spoke about previously and also through making the most for big data. If a manufacturing company is equipped with the various knowledge gained from big data, they're going be able to improve the quality of their output and production and also be able reduce their waste they create by the process. These are the key elements that can be found in a competitive market of today and can benefit any company that decides to make use of the data to gain an edge and increase profits.

Due to the numerous benefits of big data and the information being provided increasingly, manufacturing companies are moving to an analytics-based approach as soon as is possible. This means that they can resolve some of their biggest problems, as well as the small ones that pop up faster and make business decisions that are more flexible.

It is also possible for retailers to make use of this huge information. Imagine how much information that a retailer can collect into account about its customers which includes addresses, names as well as demographics and other information when a consumer purchases something or completes surveys. If the business in question can focus on this data in the appropriate way and utilizes the information they collect the data could be a treasure trove that will help the company to know how to make their customers more satisfied and enable them to increase their profit.

Customer relationships are going to be vital for the retail industry as well as one of the most effective methods for businesses to handle this is to efficiently manage their massive data. Retailers must be aware of the most effective way they can reach out to their customers, and the most efficient way they can manage the many transactions they are involved in and also some of the most effective strategies available to re-establish any business that might have been shut down due to one or the other reason. Big data is the best method to be employed to manage all of these.

With this information in mind, we have to spend some time look at the ways that big data can be used to benefit your business. Before you can learn about the ways that big data can be put to benefit your company first, it is necessary to determine where we are in the process of collecting this huge data. The sources that provide all of these big data tend to be classified into one of the three types. Naturally, a company can gather data and large data from any categories if they choose however, any information they keep can be classified into one of the categories. The three major categories of big data comprise:

stream data. The primary source of big data your business will utilize is often referred to as streaming data. This type of data includes any information that enters your IT system via the internet or connected device, and this usually is the IoT. Then, you can examine the data as it begins to surface, using it to make decisions about the information you wish to keep, as well as the data you should not keep and the data that will need further analysis of it.

Data from social media: A lot of businesses are discovering that the data they get from social media is extremely appealing particularly when they're conducting their marketing campaigns as well as their support functions and sales. The majority of the time, data will be semi-structured and unstructured, which means it will present a unique problem when compared to other kinds of data. With the growing amount of information available online, as well as increasing numbers of customers using social media, it's logical to utilize this type of data to boost your company.

Sources that are freely available. There's plenty of data accessible via free data resources. It includes CIA World Facebook, data.gov and other sources. As a business it's a great idea to make use of these resources and seek out more information will help you make informed decisions in the future.

After you've managed to search and figure out the possible sources of your data, there will be some decisions an organization will need to make in order to assist users utilize the information in the most efficient method. Some of the factors you must consider include:

How to organize and organize the data. Although storage was a major problem for many businesses only a few years ago however, there are many options at a lower cost that will manage your data regardless of how much information you've got, if this is the approach your business is looking to implement.

What is the amount of information is needed to be analyzed. There are some companies who will analyze all the information they gather. This is achievable because of the vast amount of technology that's available. However, for those on the go, would like to cut through any noise simply cannot manage all the data due to some reason or another it is up to them to decide how much and what kinds of data they'll examine.

What you will do to utilize these knowledge. The more data you can gather to use the greater confidence you will bring to any decision for the company. It is always beneficial to put a strategy to follow after keeping a lot of data to aid in making the right business decision.

Each business will benefit by making use of big data to meet their requirements. They'll be able to combine all of the information and figure out how to make the most beneficial business decisions based upon the data that is available. The last step companies should take into consideration when trying to make large data be useful is research. They must conduct a little small amount of study on various technologies that can help them make the most of their dataand assist in the analysis. A few of the options you might consider for this stage include:

Find storage that is able to handle the amount of data you've got, and also is cost-effective for your storage.

Processors that are speedier

A cost-effective and open-sourced platform for big data to make use of.

Cloud computing, along with other tools that are flexible and can assist with arrangements for the allocation.

Highly efficient throughput, high-speed connectivity large grids, virtualization MPP, clustering, as well as parallel processing.

The use of big data should be something every business must consider to use. There's an abundance of information available from customers , as well as other sources. It is certainly worthwhile for companies to examine this data, and then figure out what they will do to make use of it to benefit their business.

Chapter 3: You Are an AI Business, And Don't Even Know It! Common AI Tools You Are Already Using

The next area we'll be spending the majority of our time dealing with AI within our businesses and some of the most common techniques and tools that we have already implemented to work with AI. AI is more widespread than most business owners and even consumers realize. We frequently use products with machine learning, as well as other aspects from AI in them but we aren't aware of it.

These tools will truly help us stay safe, make life simpler, and assist us in running our business, but we've not thought about it and think about the impact, and how it's helping us. With that in mind we'll take a review of some AI tools your business is probably already using without even realizing. Also, how you can utilize these to simplify your life and still provide an excellent customer experience every time.

Tools for Cybersecurity

The fact that a lot of clients and businesses also are faced with is that cyber-attacks are becoming more frequent. The majority of businesses, however hard they attempt, aren't going to be ready to deal with all this. In addition to the inexperience at the corporate level as well, the workforce for cybersecurity will struggle with regards to being able to meet the needs of companies that will be looking to make use of this. It is predicted that by 2021 there will be at the very least 3.5 million cybersecurity jobs which are not filled across the world.

What that means is that experts in this area or even companies that have difficulty finding an expert to collaborate with must compensate for the empty slots. They can work longer but there's only so many hours in a day or through the use of technology such as artificial intelligence, they can help keep the information secure and safe for users.

In the current state cybersecurity is experiencing today's world, the inclusion of AI systems in the mix can be a major pivotal moment. They will have a number of advantages that will assist experts in this field to take on cyber-attacks and be sure that the company is protected simultaneously even when there is an absence of experts to keep up with all this.

A lot of the algorithms for AI which are employed in cybersecurity use machine learning as it allows to improve to changes in the course of time. This is beneficial since a large portion of newer cyber-attacks and malware will be difficult to detect using many of the standard techniques available. The threats will alter over time, and so certain of the innovative methods that are based on AI could be the best option to employ here. Cybersecurity tools that can use machine learning could make use of the vast amount of data from the previous cyber-attacks to assist them in recognizing and then respond to the most recent attack. Although the latest attack isn't as sophisticated but it will be similar to the previous one (this helps to save time and programming for hackers) and the machine-learning algorithm can pick up the signs of.

Another advantage we can expect to experience by making use of the AI system to assist in this area is that these systems and algorithms can let a lot of time for tech professionals. The AI system will step into play and help identify easy threats and attacks. Because the majority of threats that are detected by your system will have a straightforward solution The systems will be able to resolve the problem on its on its own, freeing up time and energy of your employees.

We must now take a look at the ways cybersecurity experts are able to make use of AI. They'll accomplish this by relying on intelligent automation to raise some danger red flags which a normal person might not have the time and resources for. As per Steve

Grobman, the CTO at McAfee that these AI technology isn't likely make human beings in the field obsolete. But they will cut down on the number of people that employees are required and still protect the company and their clients. As there is likely to be a huge shortage of people in this area, and the professionals who can do the job and reduce the number of staff required through the AI technology could have a significant impact.

Another way this type of system will assist in cybersecurity is that it will be a long time before it can categorize attacks according to what it thinks is an attack's threat. There is still quite much work to be completed in the near future, it is a good start (There is a majority of 52 percent of experts in the field who believe the AI system isn't accurate with regards to threat levels) Deep learning is able to assist in learning more about how to help us recognize the type of threat approaching us as an attack.

If you're employing an AI system for handling the cybersecurity threat, whether because your team isn't large enough to manage the threat on its own, or simply to save money or time, you could have the system handle all threats independently. If this is the case it will be based with a pre-defined process or playbook. In contrast to the varying nature caused by humans doing the work, and a lot of the errors that come when humans perform the job, an AI system isn't likely to make the same mistakes in the course of their own work. This means that every danger that comes into your system will be handled in the most effective and efficient technique that is feasible.

Although there are numerous advantages of making use of these systems to safeguard your business or your clients, as well as the information your customers give you, it is important to keep in mind that there will be certain limitations to how the technology of AI can perform in this particular field. It is able to do amazing things to help you stay safe, but it is essential to collaborate with the human-machine team to tackle some of the more difficult issues associated with cybersecurity. If you try to transfer everything to an AI-powered system will be just as risky as not having anyone in any way taking care about your security.

Artificial Intelligence is likely to be an excellent method to manage some of the various aspects of cybersecurity that are a part of your business. Every business is responsible to their clients to protect their information and ensure that their data is secure however, there always are hackers out there who would like to get access to the data and create chaos. If your cybersecurity team is struggling to keep up however it's difficult to keep up, and you're having trouble finding people willing to help with some of the tasks It is usually beneficial to use AI systems to to fill the gaps and to boost the work your IT professionals within IT IT department are also doing.

This isn't meant to be a tool that takes over what your IT experts can accomplish. It can, however, collaborate with them to cover areas where they might be unable to cover, and limit certain tasks only a few experts can handle by themselves and often detects certain dangers and attacks that human error could have missed.

Speech recognition and voice

One good illustration to show how AI is used in the workplace is the use of speech recognition and voice. Are you using Alexa or another similar product that uses voice recognition to search for things, switch on music, and even answer to a question? Have you got Amazon installed with some orders you have to make frequently and you can simply place your voice? Do you use your mobile to ask for directions or look up an address? If you've ever performed anything similar to this, or used a voice recognition software that are able to accomplish tasks using just your voice then you've worked on an instance of AI in this way.

Speech recognition will use artificial intelligence to aid in machine learning. The programmers will devote some time to making a system quite adept at recognising words as well as some of the speech patterns available. It can be achieved by using numerous recordings of voices and speech input to the algorithm over an extended period of time. Once this is accomplished it is capable of comprehending a great deal of the phrases and queries you can use to it.

Of course, there could remain a curve of learning when it comes to this. There are often various dialects of a particular spoken language, various speech patterns, and various words that are used even in the exact same language. It may take the system some time to adapt and become accustomed to the user. With a bit of practice, and the advancements that are being developed in artificial intelligence the systems are becoming more efficient every day.

Antispam filters

When we think about our emails we must ensure that the majority of the junk is removed to leave us with only the ones that matter most to us. If we didn't to happen, spam would begin to take over our email, and we'd never be able to locate the emails which were important and we would like to go through. If it's related to our business, this could be a very bad thing and could make it difficult for us to discern the positive emails sent by distributors, customers and others, as well as the spam that is not so good.

Being able to efficiently filter out the spam emails that you get on a regular basis and sort it out to a different folder while making sure that your normal emails, those that you really want to read, will be delivered to your email inbox is a problem that artificial intelligence has the ability to assist with. There are a lot of email providers that are making use of AI and

Machine Learning to make with this, and it guarantees that we will be capable of receiving the messages we need with no unnecessary extras.

This is a fantastic illustration of how machine learning is able to work. The algorithm will be presented with a variety of spam messages to look through before recognizing. By doing this it will be able discover what the most frequently used word choices and phrases that are found in spam, and also certain emails and the types that could be used to send spam. It is hoped that after enough spam messages have been sent through the system, it can be able to accurately sorting out spam from the normal emails.

Naturally, at the beginning the spam filter may not be as accurate as we'd prefer. Sometimes, we discover that great emails end up in our spam file and the spam mail does pass through. But as time passes and we become accustomed our algorithm and the algorithm gains more and more refined the filter will become betterand emails will be filtered exactly where they are supposed to be.

Machine-based translators of languages

A machine translation service will be an internet-based service, or an app which makes use of some of the techniques that machine learning provides to translate large amounts of text into and from certain other languages available. The application will begin with a text source, or the text you originally wrote that you wish to translate and convert it into the second language or the language you want to translate into.

While, some of concepts associated from this tech, as well as the interfaces that are built into it are fairly easy to comprehend. The science behind it and the work to be completed will be complicated and bring together various elements like web APIs semantics, cloud computing deep learning, to name just only a few.

Since the early 2010s the introduction of a new artificial intelligence technology that is known as deep neural network, has enabled speech recognition to achieve the level of quality that clients are now accustomed to. Speech recognition technology is among the key components involved in translating speech and text from one language to another.

We must begin by looking at the way in which translating of texts is likely to occur. What is the best way for a machine to read a sentence in one language, quickly and accurately translate it into an entirely different language. This kind of translation is possible in a variety of languages to provide the user with the option of choosing. Although each program might be different, we're going to look at the way that Microsoft Translator works to give us an idea of what it does and aid us in understanding how some of the translation tools are available to us will function when used for business.

When it comes to the use of text translation There will be two different types of technology which can be utilized. They are the NMT or Neural Machine Translation and the SMT which is the Statistical Machine Translation. This NMT is the newest version of this technology, however they are both able to work to provide the results we're seeking when translating different texts.

The development of Microsoft Translator using SMT technology is based on more than 10 years of study on natural learning which Microsoft has done. Instead of having programmers create and create rules they created themselves to get the language translated the modern translation systems are likely to view the issue of understanding how the translation of text between the two languages occurs. This will be accomplished using a variety of methods that include some of the current translations they've gotten taken from human beings and employing a variety machines learning techniques as well as statistics.

The parallel corpora is likely to be a bit like it's Rosetta Stone when it comes to translation of text. This means it assists the algorithm learn things such as idiomatic translations, phrases words, phrases, and so on in the context of a variety of different languages that are commonly employed in conjunction. Then, we use modeling techniques from statistical analysis and effective machine learning algorithms to aid the computer in solving any problems it encounters when it is deploying these translations.

This new method that is expected to benefit because it's not dependent on grammatical rules or dictionaries that would consume many hours and have too many rules of grammar to think about between various languages Artificial Intelligence translations will help to provide the best translations using a machine that doesn't require any human interaction.

The process will take place by using the neural network. The system is expected be learning in a way that is comparable to humans' brain. It will be able to detect patterns that are visible within the different levels of the documents, and will strengthen its understanding when it is able to answer the question correctly. The more often the translation service is able to do this, the better it will become.

There are several processes that need to be taken to provide an automated translation service, with the aid of the neural network. They will include:

Each word also referred to by the name of the 500-dimension vector which represents the word, will be traversed through its initial layer. This layer will encode the word into an 1000-dimension vector. This vector will be used to define the word and what it signifies in relation to the context it is when compared with the other words in the sentence.

When the system is capable of recognizing each word in its own then encode the word into vectors, the same procedure will be repeated a couple of times. This procedure ensures that each layer can refine to ensure that we comprehend the meaning of each word in the entire sentence, not just as a standalone.

In the matrix we will use to create our final output we will be in a position to use this output and incorporate it in the layer of attention. It will make use of the different calculations made in order to remove any words that are not utilized or are not needed in the target or the second layer.

The decoder layer, also known as the layer that is translating, in turn capable of translating the word you have selected or the phrase typically in the language you select. The output from the final layer could be transmitted back into our layer of attention. This is to make sure that we know what word will be the next word to be translated.

Security surveillance

Artificial intelligence will be a major factor in the field of security surveillance for people as well as for different corporations. This will allow us to use computers and other machines to analyse the footage it observes without having someone present to perform the job. This is a great thing in terms of security of the public and will assist first responders as well as police officers to identify incidents and criminals. There are plenty of scientific and industrial applications.

Many businesses will depend on a variety of surveillance cameras. They can use it to keep track of thefts and other problems happening in the shop. It's a useful item to keep in the store for security of customers. Have you ever entered an establishment and seen the security camera and not thought what you would do? We all believe that they are there and we've become used to them. We know that they are there to protect people and to ensure we're secure.

When using these devices, one has to go through the endless video that are available to determine if somebody is taking something or taking risks to others who are around them. The process is usually completed after the fact , too after the police or other authorities have been involved. The majority of companies don't want to hire someone to look at the video all the time, so the task is completed following the event. If there's no suspicion of theft or risk to the clients the all the hours of recording is not considered because it takes too long and takes too much effort.

The quantity of events that happen within the company and not being noticed by employees and customers which could result in a loss for the business is not known. The information could be captured in the video, however the majority of companies aren't prepared to deal with watching all the videos, which means that it is not noticed. Instead of hiring someone to stay in a room all day and watch the film, and often overlook things because they glance away or get bored or aren't paying to pay attention and don't pay attention, we can employ artificial intelligence to complete the job.

Through video surveillance, you'll be able to train the artificial intelligence algorithm what you would like it to behave and then show it the actions which must take place so that it is able to detect when violence, theft or other unacceptable behaviors occur. Most often, this is accomplished through the presentation of videos and images for the machine until it begins to detect when these behaviors are taking place or not.

The system is designed to detect in the video that something isn't right it will send an alert or another signal depending on the information you input into this. This helps companies become more efficient, ensuring that thefts and other crimes will not happen more often and help the company to maintain cash at the till without the need to constantly look through the hours of footage which is available.

The tools that make use of AI are increasing each day. These tools are fantastic ways to aid us in working on improving the systems we already have, and making sure that the business we run runs according to our expectations. Most of the time, we work using these types of tools and not even realizing they exist, and we actually work using them.

Chapter 4: AI and How It Is Used in Retail

We've spent some time to research the basics of what AI can do and the ways it could be applied, it's time to look at the ways in which AI can be used in the different industries we might operate our businesses in. There are many various industries and businesses that can gain in the field of artificial intelligence. This guidebook has spent time looking at which ones and the ways they can benefit from the technology. For this section we will take a closer look at how retailers and companies can rely on AI to aid them.

It is a fact it, the way we live has changed, and we've progressed a lot since the days when people had to go to a brick and mortar shop to buy anything they desired. Nowadays, it's possible to purchase almost everything that you need by going to an online shop due to the technological revolution that has taken place around us in the last 25 years. It's gotten to the point that it's even in a position to influence the way the customers will display in physical stores too.

In light of all these changes and developments, retail stores will gain a lot when they begin introducing the use of Artificial Intelligence, and the technology that goes with it in their stores. AI is thought to be one of the foundations for technological advancements across the world of business, and many businesses, regardless of what industry they're in have been successfully benefiting from the benefits of customisation automation, recommendation, and automation engines that work to ensure that the company has more customers.

This is the case when we look at the retail industry. AI is beginning to play an even greater role than it has ever before in relation to the shopping experience since the consumer is capable of shopping around and discover the best products they'd like to use according to their preferences. Furthermore, VR, also known as virtual reality, is set make shopping even more convenient to shop for those who want to test out an item without having to put it on or wear it on first.

Another aspect that many retailers are focused on is a specific type of technology called RPA (also known as robotic process automation). This is beneficial to businesses, but can also help the customer as it's designed to get the appropriate products to the appropriate people. Another platform you can use is called WorkFusion which assists companies in improving how the shopping experience goes for their customers.

As you've already seen that there are plenty of possibilities for a business to make improvements and increase sales through the use of AI and the tools that are associated with it. To assist us in understanding more precisely how retail stores can utilize AI to grow and make more money and increase profits, we will take a look at four examples of AI that have already been discovered within the retail sector. These include:

Browse through catalogs online

The very first thing we can observe through AI for retail sector is the capability of the buyer to peruse an online catalog to view what's available and to purchase items from the catalog. What number of times we walked to an online store that is one of your preferred stores and looked to see what they had and what new products they offered, and much more? Even when you went to the store without the intention of buying something in the first place and all those vibrant options as well as the convenience of shopping online caused you to commit to purchases, which led to an easy-to-make gain for the company.

AI has created various suggestion engines, which are more sophisticated and assist customers in finding every item they're looking for. It is usually completed in less than the time it's going to require with the older systems in use, or with the time of the customer going to the store to browse.

Instead of having to visit the store to browse through hundreds of items and physically, consumers can browse through a catalog online which will inquire about a variety of questions regarding the products they are looking for, such as the dimensions, personal details as well as the color preference and other details in relation to the items they want The catalog is then created to showcase certain items that may be appropriate for the individual.

It can be done with any kind of product that you want to use it for, obviously. The above example is most likely for clothing or shoes, however it is possible to use this type of catalog online to assist with things that aren't apparel related, such as accessories, music books, and other items. The idea behind this is to facilitate shopping for the consumer by speeding the process, making it easier rather than a hassle which results in higher profit for the business.

Advanced gesture recognition

The second feature of AI which is readily available for use in retail is the concept for advanced touch recognition. Alongside having the benefit of certain catalogs online mentioned in the previous paragraph, it is also possible for retailers to check in and see the success of a new product is, or even a particular product they're trying to sell. They can see the reactions of their customers to the item by observing the hand movements and expressions of the buyer as they interact with the product.

The way in which a buyer or shopper will react to an product, particularly when it is compared with other items, is going inform the sales associates (which are robots to do this) which of it is an item that is deemed to be to be a dud or even a top-selling product. Consider how this could benefit your business if you were in a position to determine which products customers are really interested in and receive actual feedback from your customers with regards to a brand new product you're thinking of launching but aren't certain whether it will work away or be rejected.

In addition to the hand and facial gestures, and the advantages associated with them and the benefits that come with it, when a store can learn more about how a user is likely to react to a particular item, the recommendation engines will be able to make use of the data. They will perform in a more efficient way to find the item that is perfect for every person, based on what they've bought in the past, what they tend to be searching for and how they've react to various products that were displayed during this test phase using AI.

Take a look using an online mirror

This could be a fantastic alternative to partner with many clothing retailers on the internet. There are times when people want to buy an outfit, a dress, a business suit, or even something else to attend a major event however they may not live near an enormous store or big city, or do not have the time to walk from one shop to another to locate the items they need. Sometimes, the item they want is easily found on the internet. They require the item, but aren't sure if they should buy the item due to the uncertainty of whether the size will fit or if the color will fit their body, or whether this is the best design for them.

Instead of going to a physical store to try on clothes on the spot, placing an order and hoping the product is the correct size and having to contend with returns and more if the garment doesn't fit and isn't what you expected, it's now possible for retailers to use virtual mirrors. This allows customers to test various items and determine if they like the items, all in a short period of time. This can lead to more satisfaction for the customer, and ensures that the purchase will turn out to be exactly what the buyer would like, and will save the customer a lot of time and headaches with regards to returns and refunds.

If the customer is searching for the clothes they require, but aren't sure if it will fit them well or is the type they are looking for or would like to know what other styles of clothes or accessories or shoes would match the outfit the virtual mirror could help them with this. It is possible to do it from the comfort of the client's at home (if the business offers it and the client has access to the mirror) and can make it easier to save time and money, while also giving the client exactly what they need.

Video analytics can be helpful in customer engagement

Another way retailers can interact with AI and the methods that are associated with it's to use video analytics. Naturally, retail stores are likely to have cameras within their stores. They are likely to be utilized in a way to ensure that the stresses are as secure as they can, as well as aiding in customer service and to ensure sure employees are acting in a manner that is appropriate and offering high-quality service.

As you can see, there are numerous choices available with regards to the things you can accomplish with your camera system but why not make use AI tools to boost your customer interaction with these and also to make sure you provide the kind of service and products your customers want and help you gain an advantage over your competitors.

The various technological advancements in the field of computer vision today will lead to an environment that's more enjoyable for customers and can make more money for consumers when used correctly. This type of surveillance will benefit retailers since it gives information on the level of exposure a buyer has to the product, how engaged the consumer, and the way that customers take throughout the store. This is crucial in assisting the retailer placing items where they are needed for the highest sales.

It could be an issue that requires a lengthy time for retailers to develop and implement on their own. But an AI system would be able to process everything and return with useful data, as well as an efficient plan for navigation in the store that is based on the buying habits of your customers, what directions they used while in the store, as well as the products they used in the store. Imagine the impact they could bring to increase sales for your business if moving a few things could make them more appealing to many customers, and also made customers more inclined to complete that purchase? That's definitely something AI can accomplish by using the video cameras you already have inside your shop.

As this chapter was in a position to inform us before that there are plenty of instances when we'll be able to use artificial intelligence, including within the retail industry. Although many believe that this is one of the last areas to see AI or machine-learning will ever appear and that only fields such as science and technology will benefit from it, there are plenty of possibilities for retailers to make use of this technology. We have discussed four major options that are easy for retailers to use to boost profit.

Given the satisfaction that retailers have had the chance to enjoy by incorporating AI in their stores, the technology is expected to play a greater part as time passes and soon will be an integral part of the retail industry. If your retail business is looking to work with one of these AI options and looking at the other options available to in the future it is possible that consulting with an IT company will ensure the you as well as your team know which strategy and what types of AI will assist you in selling more products and earn more profit while making sure clients are pleased and content and.

Consider all the ways AI as well as diverse technologies that accompany AI can benefit your company. By helping you enhance the customer experience by allowing them to customize and ease of use, this results in a customer that is more satisfied and makes large purchases. So, in the long run, you'll see a better profits. This is why the majority of companies, even retailers are shifting to artificial intelligence.

Chapter 5: Bringing AI Into Your Marketing

Another area that stands to profit from the application of AI will be marketing. Whatever industry you're involved in, your marketing will to play a major role of the overall process. It allows you to connect with customers and place your business prominently in their minds. It assists you in talking about the latest products that your consumers might be interested in. It assures you to build the most favorable image for yourself by doing so.

In the realm of marketing there are plenty of different strategies you can utilize to get in touch with the people who are most important to you. You can use social media, you can make viral videos, advertise on TV as well as on radio stations or in magazines and even in newspapers. Each of these options will make it much easier to connect with the client. However, the most important element that can take you to the top of your game is likely turn out to be AI.

There are a variety of methods you can choose to make use of AI and the tools that it offers for marketing you conduct, regardless of what type of company you operate or what type of marketing strategy you wish to implement. A few examples of what you're capable of doing when you are running a marketing campaign by using AI include:

Deep learning

The first thing we should be looking at in this regard is the notion that deep learning is a concept. Deep learning is regarded as the research that will to be the foundation for much of the marketing we encounter in the modern world. This type of learning will help a computer learn a wide range of abilities, such as understanding photos or speech as well as text. It can help the system in learning to provide answers, answer every question that it receives and provide suggestions that are specific to the user.

Let's look at an illustration of how this can be done. It is evident that Facebook is working with deep learning in numerous instances to improve the filtering of posts and the types of ads you are exposed to. Facebook can look at the content you respond with the greatest, what you've done online searches for and more, to ensure that the content that you see in your profile is customized to the most important things for you.

Content curation and various suggestions

This is a common feature that you're sure to appreciate when it comes time to provide helpful suggestions to your customers to help them increase the value of their purchases. You've probably seen numerous websites with a message similar to "If you like this, you could like this too." ..." what you like." What people might not be aware of with such recommendations, however it is that they've been built on deep learning and other functions of AI. Machine learning will examine a variety of information that is available being released about the actions of people, customers generally, to get an accurate idea of what the individual is likely to perform next or to enjoy the most depending on the information they've have looked at.

This adds a bit of personalization and keeps the client there. The more that you can get your potential customers to click the higher the likelihood that they will add the item to their carts and then make the purchase once they're done.

SEO optimized headlines

It is also known as clickbait. It is important to ensure that the headlines you're giving your customers are appealing, grab their attention and will likely get them to click through your site. If a customer never visits your site to explore and read the content, you're unlikely to achieve the sales you'd like. However, we are aware that the majority of headlines used by clickbait are designed at the very least.

The good news for marketing professionals in all fields, and especially those who are among those struggling with writer's block is that researchers in Norway are learning to utilize neural networks, deep-learning and neural networks to create headlines automatically for you. This means you'll will have the best headline that will attract your clients every time, regardless of whether you're struggling to come with the best one for your specific needs.

Product suggestions

Naturally, in the field of marketing, content isn't likely to be the sole location that you can utilize the features of deep learning to provide suggestions. If you decide to browse the online store of many diverse businesses, such as places such as Amazon You can observe the way AI is used to recommend products to customers according to what they've seen in the past and the purchase they've made.

Speech recognition

In certain instances your marketing campaign might require the abilities in speech recognition. If you're working with alternatives like Alexa, Siri, or Cortana on your smartphone or tablet, then you're well aware of how far technology has improved in the past few years. This is getting more essential in the world of marketing since it could aid in SEO. This is because you must to ensure that the information you post is designed for search queries that aren't just written out, but in question form, which can change your game in how you will create your social media platforms, your blog, and even your personal website in the process.

Search

One of the most impressive instances for AI in the world of marketing will be the search feature. The queries that are made use of can direct the user directly onto your Facebook pages or website. Once they're there, it's more likdly that the person will make a purchase of some sort.

An excellent example one of these is Google employing RankBrain. This is a kind of machine learning technology which is capable of analyzing the spoken and written search queries, and later transform the results into search results that coincide with what you are looking for. It goes beyond just returning the words and keywords you would like to search for, as it will examine each query that it receives with other similar queries, and will return ones that are most likely to match the exact thing you're trying to find, as well as ads that are more pertinent to the customer's needs.

Ads targeted at

The next aspect we should consider when making use of artificial intelligence in your marketing strategy is targeting your ads. Many websites, including Facebook and Google are likely to utilize small pieces of code to monitor the number of visitors who visit their sitesand tailor the ads they offer to them in accordance with the visitor's location. As of now, this procedure isn't the best and difficult to use but it's likely that major modifications will be made to this process over time.

In the present, Google is beginning to conduct some research and experiment with AI technology. They are making use of RNN, also known as Recurrent neural networks, which could help them "remember" details for a brief period of time. This could benefit the business since it removes the requirement for software which can perform specialized ads and help make this method of marketing which is more effective in general.

Chatbots

As the options in the marketing channel start to expand and evolve marketers must be sure that they handle all interactions in a quick and cost-effective way. It's not going to be cost-effective if a large companies have to join and manage all social media websites, responding when someone responds. They might need a team just to do the work of social media on their own, which can cost a lot more than the other services the company is spending on in the marketing campaign.

Instead of having entire teams working on social media or on other channels of marketing, such as the web, the business could use chatbots. This will be an artificial intelligence that will learn to communicate with human beings. It won't be able to do everything, so having an employee to take care of some tasks may be beneficial however, it should be able to manage some of the more routine tasks like answering frequently asked questions or helping with purchases and even arranging travel in certain instances.

Marketing automation

The business may also decide to utilize the term marketing automation. This will utilize AI to aid in engaging customers. In this phase, the system will be able to analyse the habits of the client and provide the content specifically tailored to the person. The purpose of making this happen is to to move the client through the sales funnel and whether the system can suggest content, interact with them on different platforms or send them an individual email campaign. This allows your company to connect with the customer wherever they are, by offering them exclusive information and content to meet their needs and increasing the odds that they will purchase.

Dynamic pricing

The final option we will examine in the incorporation of AI into the marketing process is to use dynamic pricing. This kind of pricing is frequently used in traveling, however it is also a possibility to use in other areas if the company chooses. This type of pricing is the place where the company is able to determine prices for the products in accordance with demand and availability at the moment.

There are plenty of businesses that could benefit from this. Consider the time you are shopping on Amazon and put something in your cart to contemplate for a while. If you notice that the price fluctuated either way or down during that time frame You've seen the effects of dynamic pricing. This allows companies to adjust prices according to the market's demand and helps them increase profits overall for the same item. If there is a that the item is in greater demand, the dynamic pricing likely to increase the cost for those customers. However, if the demand is lower prices can be reduced also. The AI algorithm can determine this period of time for the company, which makes it simpler to make changes without having to monitor the situation with the same exactness.

While we typically don't even think about the use of AI in marketing however, there are lots of ways this type of system are capable of implementing and change the method by which companies communicate with their customers. Utilizing even one of these strategies will give you an advantage over your competitors and assist your business to expand.

Chapter 6: Can Healthcare Benefit with AI?

The next area we'll be looking at in the direction of how AI could improve an industry is the realm of healthcare. In the case of our health, specifically in areas that involve vital to our lives, AI can really come in to help and offer us various methods to make doctors more efficient. Although there is much to be done before we reach the point of healthcare that relies upon AI as well as systems that rely on AI especially in the area regarding data privacy and worries about care being not properly managed due to the error of machines and lack of human oversight the system, there is lots of possibilities that health care organizations, technology companies and government agencies might be willing to experiment with the tools and solutions available and examine their effectiveness for the health care sector.

With this in mind we might want to look at several ways in which AI can aid the medical field as well as certain areas that have lots of scope to see it improve in the future. A few of the most well-known and well-thought-out examples in AI that could occur in the realm of health care include:

AI-assisted surgery

The first aspect of the field of health care that we will investigate in terms of the ways we can utilize AI technology will be robotic surgery which is assisted by AI. With a market value estimated to be $40 billion in the field of healthcare robots can use some of the information available from pre-operative medical records and use it to guide the instruments that surgeons use during the procedure. This can actually help decrease the length of time that patients stay in the hospital, which is currently 21 percent.

There are many things to look forward to in the field of surgery which is performed by robots. For starters, this type of procedure is thought as minimally invasive meaning that patients can heal quicker and have a smaller scar. Through AI robotics, robots will be able to collect the information from previous operations and then help develop new techniques for surgery to improve the outcome. The results are extremely positive and promising currently.

A study was conducted, 379 orthopedic patients analyzed. It was discovered that when they went through an operation that was assisted by AI the patients experienced five times less problems when compared to patients who performed the surgery by themselves. In a different study, a robotic was selected to perform eyes surgery, for the very first time and the most sophisticated surgical robot, called the Da Vinci, allows the surgeon to perform procedures that are extremely complicated, but with more control when than other methods that are utilized regularly.

Another instance of how robotic surgery robots that are AI-controlled can aid can be seen in cases like Heartlander which is a tiny robot that can assist heart surgeons. The tiny robot is able to be inserted into a smaller cut that surgeons place in the chest for the mapping as well as some treatment on the surface of the heart. This makes the procedure less invasive and more efficient, and helps the patient recover and heal quickly.

Virtual nursing assistants are nurses who work remotely.

The next thing we're going to consider when we consider adding more AI in the medical field is the possibility that we can utilize this technology to provide more virtual nurses. From being able to communicate with patients who are brought in, to directing patients to get to the facility that is most effective for them, it's possible that the nurses who are virtual and can function with the help of AI are in a position to save this medical field $29 billion per year as well!.

Virtual nurses are accessible all day, every day This makes it easy for them to respond to any questions, keep track of how patients are performing and provide some quick answers that are required. This makes them more effective than human nurses and may assist in covering a little when shifts are not full on staff.

The majority of applications for these virtual nursing assistants will facilitate continuous and frequent communication between the health professional and patient, in addition to during their appointments. This alone can help stop the re-admission of patients to hospitals and decrease the amount of hospital visits that aren't necessary. In certain instances like with care angel, for instance Care Angel virtual nurse assistant you can make use of the tools of AI as well as your own voice to perform an assessment of your health.

Assistance in diagnosing and clinical judgement

AI can also be able to help diagnose certain patients who are present in the clinic. This is admittedly an aspect of AI that is relatively new and isn't much of research done in this area, but there are a few instances which are very interesting and demonstrate what it takes to make this perform.

In the beginning, a research study carried out in Stanford University tested the algorithms of AI in terms of how well they could identify skin cancers compared to dermatologists. In this particular study they were able to perform the same as the human doctor.

Another study was conducted in collaboration with an AI software company based in Denmark that tested some of the programs they used using deep learning. This was done by making the computer listen when dispatchers received emergency calls. The algorithm was able to study what the caller was saying, the tone of voice employed as well as the background noise background and whether an cardiac arrest was taking place with an efficiency of 93. The rate of success for the same situations for human dispatchers was 73 percent.

As per Baidu Research, they are at the point of results to conduct preliminary testing of an algorithm that they developed to use deep-learning. The algorithm they have chosen is intended to be tested to determine whether it's able to outperform humans in identifying metastasis in breast cancer. Prime Minister Theresa May announced a new technological advancement in AI which is designed to assist to improve the UK medical system, known as the National Health Service, to assist in predicting cancer at certain stages earlier and help prevent hundreds of cancer deaths by 2033.

You can clearly see that even though these methods aren't employed all the time, as of moment however, there are plenty of algorithms available that are being studied and could be used to determine the type of disease patients are suffering from. Based on the situation the algorithms are equally efficient, if not even more efficient than their human counterparts. If the algorithms continue to improve and improve and become more accepted as time passes, then it is possible that they will save a lot of lives in the coming years.

Tasks related to workflow and administration

The next aspect you can accomplish with the aid of AI tools will be the administrative and workflow aspects. If the tools were utilized properly, it could result in saving the healthcare industry around $18 billion. This is because AI machines will be able to enter and help nurses, doctors as well as any other type health care professional reduce time spent on their jobs. It is true that time is money.

Technology, such as transcriptions that convert from text to voice, could help take the tests patients require, recommend the correct medication, and even fill notes on charts to be used by the doctor. When a doctor can complete these tasks faster, without having to sit and complete these things on their own or hiring someone to complete the task, this saves time and money as well as making more satisfied patients.

One instance we can examine when we think of making use of AI tools to aid in admin tasks includes the well-known partnership with IBM as well as Cleveland Clinic. Cleveland Clinic. The Cleveland Clinic has been working in conjunction with IBM's Watson to collect large amounts of data and make sure that doctors are to be able to offer the patients with treatment options that are more effective and more personalized during the procedure. One way Watson Watson program can use to assist doctors is that it's likely to absorb thousands of medical documents using natural language processing. It will use the information to create treatment plans that are supported by scientific research and are much more well-informed than they were previously.

Image analysis

There are tons of images coming in at the providers. They will have to examine things like ultrasounds, X-rays and brain imaging, among other. Each patient might require one or more of these tests and they must be completed in a timely and efficiently. If a doctor is treating 50 patients that have had their images taken in the last week and need to review the results and discuss the findings with their patient it's not surprising that it can take much time for the human caregivers and could create a lot of overwork during the procedure.

A research team which was headed by MIT has managed to create an algorithm based on machine learning that's capable of looking at 3D scans and analyze them 1000 times faster than the human doctor can perform. This analysis is almost real-time, which can be very helpful to surgeons while they are operating and can may even speed up the tasks a physician is able to perform. It's not going be able to substitute the traditional doctor but it will to provide a new dimension that allows doctors perform their work more effectively.

Additionally there is the expectation that AI will aid in improving some of the latest radiology instruments that are to be made available, tools that can assist in diagnosing problems without having to depend on tissues all the time. These AI imaging analyses will be useful in a less remote region. They might not be able to be as accessible to medical devices or employ the same types of doctors that the ones located in bigger areas will too.

The best part is that it will make healthcare accessible to all. Telemedicine, if utilized in conjunction using AI technologies, is likely to be more efficient as patients, regardless of the location they are in can utilize the camera of their phones to send photos of things like bruises, bruises, cuts, rashes and so on. They will be able to determine if medical attention is actually needed, freeing the resources of the hospital, and also saving time for situations that aren't major and don't require to be examined by a physician.

The world of healthcare is complicated, even with the regulations and many other things which these facilities must be focused on. In addition, the use of instruments of AI to help to not replace human healthcare providers, but assist them to study data and see patterns in the genetic data of a patient. It can also assist in diagnosing quicker and provide more efficient service for the patient. If you find yourself in a position where one minute can change the course of the life you save machines learning and AI will make a huge impact on the future of healthcare as well as for all patients who are capable of using this kind of technology.

Chapter 7: AI in Telecommunications

The next area we'll take into consideration in terms of how AI operates is telecommunications. If you believe that telecommunications aren't connected to AI and data science, then you are certainly incorrect. In fact, according to the survey conducted in the Harvard Business Review in 2018 approximately 70% of the top executives of large companies believe that they've been able benefit from projects that include data science and AI within these. Now we are at the point that data science is required to be present. It was previously the method companies used to establish a competitive advantage over their competitors. In the present, if you do not incorporate this into your business, particularly in the field of telecommuting it will leave you to fend for yourself.

AI is currently used in many different sectors, and it will be a major factor in the business of telecom. The most innovative operators within this sector will utilize artificial intelligence and machine learning to aid in various ways, based on how they want to assist their clients, as well as the outcomes they want to achieve. For instance telecom companies could make use of AI to improve the processes they're employing to make more money and increase satisfaction and retention of their clients, and ensure that the reliability for their system is improved.

There are three major areas in which telecom companies can profit the most when they make use of AI and other aspects that comprise machine learning. The three areas of focus include:

Service to customers and retention of clients

The first aspect we'll explore in terms of how telecom companies can make use of AI is customer retention and service. Top telecom companies use the tools that are made possible by AI to assist with various types of processes. This will include streamlining methods for customer support, personalizing the offerings they offer customers, and assisting by utilizing chatbots that are automated.

There are one or two exceptions AI-powered customer support solutions are often divided into two distinct categories. One is customer service communication and the other one is about an engagement with the customer as well as a better experience for the customer. Finding a solution and being able to enhance each of these issues will help reduce costs and will make the business better overall.

The first thing examine is how a company working in telecom is able to resolve the issues that their customers face. Given the size of these companies are at the moment it is a massive amount of work for human agents. However, AI algorithms can be in the picture and manage an enormous quantity of data and interactions entirely on their own making sure that the company is able to learn more about prospective and existing customers, while also ensuring that questions and concerns of customers are addressed efficiently and in a timely way.

The customer service solutions that can be powered by AI are likely to be portrayed in a variety of ways in the world of telecom by chatbots' interfaces. But that's not always the scenario. In certain instances, the AI algorithms will exist but will be behind the scenes, helping ensure that the department responsible for customer service is most cost-effective as is possible. There are numerous instances of how the algorithms of AI could be utilized to aid large telecoms when concerns communication for customer service. These include:

1. They could serve as the intermediary between the needs of the client and a live chat or help center.

2. They can assist in directing customers and their needs to the appropriate customer service agent . They can also assist customers who call with the intention to buy to get

straight into the department of sales and reduce time.

3. They are able to help analyse the requests of customers and the data of the network, to ensure that any issue of the customer is solved in the quickest way.

4. They can assist in identifying those hot leads by reviewing all emails received by the company and routing them to the appropriate people in sales.

5. They allow customers to explore and even purchase various media items, with just their voice and spoken words instead of being dependent on remotes to get things accomplished.

6. Chatbots for entertainment are included in this and operate using the native platform the company owns or on an Facebook Messenger platform based on what's easier.

One area where AI technology will be utilized is when we wish to incorporate it into our customer support process as well as a customer support agent. An excellent example could be Ask Spectrum virtual assistant, that runs on AI and will assist customers with general queries about their account, account details, and even troubleshooting. The customer will get questions answered by the chatbot and the amount of ways in which the chatbot can accomplish this depends on what the company would like to accomplish.

The assistant is programmed to offer the user hyperlinks to Help Center and other helpful suggestions, or when the situation is complicated, it could send the user to an Live Chat representative. These are generally much faster to respond, with no having a person in front of them, and are able to resolve a majority of the problems that a customer might face. This can free up the time and energy for the Live Support staff to better respond to what the customer requires.

In the end the long run, this will benefit both sides. Customers can have answers to their questions and problems resolved in a matter of minutes, instead of waiting in a phone line more than 20 minutes. The customer service department of the company won't need to employ numerous staff members to accomplish the job also. When the client has a more complicated requirement that they aren't able to obtain assistance elsewhere but they are able to reach a live person to assist.

Experiences and sales that are personalized for the customer

Another way that AI employed by telecom companies could be effective is through boosting sales, and assisting with the personalization of user experiences. Alongside chatbots working to help in customer service as well as inquiry routing systems, it's possible that AI will be used and aid the telecom business to improve retention of their customers while earning more profits for every user who remains loyal to them.

There are many ways in which AI could be integrated into the picture and aid telecom firms, particularly in giving them a better customer experience. The strategies are ones we've discussed in this book as we've progressed. The ways in which your telecom company could profit from using AI techniques could include:

1. It is possible to provide suggestions that are customized depending on the user's needs and their preferences for content, and patterns of behavior, they have demonstrated over time.

2. It can create the best types of cross-sells and upsells deals, at the right timing and with the right customers. This is more effective than what you'd be able to observe through your sales representatives.

3. This program can allow you analyze the data and calling packages you have and figure out which will be the most effective for each type of client you're working with. If you can provide the customer with the correct

package, they'll be more likely to complete the purchase.

4. It is a way to identify and resolve any issue the client could face prior to the time that the customer realises what is happening.

5. They can examine the consumer sentiment as well as brand coverage and social media to assist them understand what drives customers to a certain company, and also what it is that are that lead these customers to leave the premises to improve the way they're doing.

An excellent example of how this can be seen is Comcast which is the largest broadcasting and cable television corporation worldwide when it comes to revenues. They've managed to introduce a brand new voice-controlled device that will make it much easier for users to use their system using natural language. This company is capable of using AI to assist them in processing all the information they can collect from their customers. Furthermore, they can make use of vision from machine learning to recommend the most effective content for their customers.

Although this can mean more profit for the company but we should also take a moment to consider what this could mean for customers. For the consumer using an agent that is AI can mean that the customer service experience will be more pleasant, rather than waiting up to 20 minutes in order to speak to a customer service representative and then having to wait for a long time to figure out what's wrong. The algorithm will help and get the issue solved in minutes, based on the nature of the issue as well as the reason customers are calling. This can increase the satisfaction of customers as well as a greater percentage of retention for the business.

Assisting with the analysis of networks and any maintenance that is predictive.

As we stated at the start of this section, telecoms are going be able to participate in two different areas of the telecom industry that is network maintenance and customer support. With the expansion of more than 40 networks, and an estimated 50 expected to come out in the near future, we are beginning to gain knowledge about the ever-growing need for data consumption in all kinds. The ability to optimize the networks you're working on, and to get them to be able to handle the greater volumes of data consumed both in the present and the future, could be one of the major decisions in strategy that the telecom industry will be relying on.

It is likely that in the case of some of the newer solutions driven by AI maintenance of networks is frequently regarded as one of them. This is because they are focused more on a method which is software-centric and likely to be as close as is possible towards self-learning, self-optimizing, as well as self-healing networks.

The past was when a network service provider was required to send certain employees to the location to check what the equipment was functioning. This was a good idea at the time but could lead to a lot of mistakes and delays and didn't do much to improve the experience customers were experiencing. There are still companies that use this approach, a lot of the unnecessary and urgent checks they conduct could be prevented if data science was added to the mix more.

Today, AI algorithms can enter the scene and watch millions of bits of data and signals to monitor the hardwarewithout having to be present to accomplish it. This will ensure that we are aware of imminent problems immediately when they start taking place, and the company can then come up with some things to correct this, based on the nature of the issue is. Sometimes, the human agent may still be required when problems arise, but usually just a restart of the software or load balancing procedure will suffice.

For us to learn more about how network analytics powered by AI will work we'll look at some examples below:

1. A system powered by AI can restart the towers solely according to the behaviour they display. If towers do not connect to the Internet, it will be able to reboot them, which could resolve the issue.

2. The algorithm can look over and determine which component of the network is likely to require more time and investment and which part is most likely to provide the most returns on its investment.

3. Additionally, network operators are capable of using AI to identify those parts of the network with the highest number of users and will be the ones to have the most benefitted from changes within the network. This aids the business in gaining greater profit.

4. It will help you enhance the results you experience with the network. This can include statistics on usage in real time and any movements that occur daily, as well as the weather data they provide.

5. These systems can improve the use of the network as well as the level of satisfaction that the client receives through the dynamic allocation of resources.

Like you'll see there are numerous ways in which an AI system can be used to make its mark in the world of telecoms and make sure that they can give the best customer service that is possible and also increase profits in the process. Many of these companies already employ at the very least a few AI systems to do this!

Chapter 8: Adding In Some AI in Finance

Another sector which has been able truly benefit from to AI will be the finance industry. AI has performed wonders in the field of finance, allowing companies to effectively deal with a wide range of aspects of their business. This type of technology will aid in not just providing the best customer service which might not have been feasible before, but will aid in improving the effectiveness for the bank. ensures a high degree in transparency and make sure that regulations are in compliance with the deadlines.

There are a variety of ways the finance industry and banking will benefit from the application of AI and the incredible tools equipped to make this happen in addition. Examples of how the banking and finance industry, along with other similar businesses can profit from AI are:

Risk assessment

The primary component that is included in AI is learning from the data taken in the past so it's only natural that the realm of AI and all the technology that is derived from it will be able to be successful and perform well in the field of financial services. Records and bookkeeping, as well as a lot of information will be essential in this type of business. Think about the various data that's processed and utilized for anything connected to finances.

Let's take a closer look at the way credit cards are utilized as well as managed and distributed. In the present, you're expected to utilize a credit score to determine whether you'll be suitable to receive one of the cards. However, the ability to classify people into who are not haves and those with credit scores isn't the most efficient approach to make use of for business. Instead, the information of the person that includes the number of credit cards they hold as well as their current loans, their habits with regards to payment and much more could be utilized to tailor the credit card's eligibility for the individual and also to alter the rate of interest permitted on the card that is most beneficial to the financial institution.

While this is beneficial for the customer and the bank, we have to think about what kind of system will actually be able to bring the pieces together and make it happen. The answer is simple: AI. Because it is all about data, and being heavily influenced and supported by it going through all of the data the financial institution is in a position to collect from their customers could provide the AI software the capability to give faster and smarter suggestions on credit offers and loans. They are also more accurate and efficient in performing this task than a skilled loan officer , too.

Artificial Intelligence and Machine Learning are set to begin taking over the position of a human analyst in a short time because mistakes in this type of field could cost millions. AI will be built on the concept of machine learning. That means the program will be able to perform the task better with time. This means that there is a less chance of making mistakes during the process. This means that, although it might require some effort to set up the program to function in the way we would like and to be installed and operational The more it is used for this type of task and the more experience it can become.

As you can see, this will help both the customer as well as the business in a variety of ways. The business will gain from this as they are in a position to get loans approved faster and help people get the cash they require and earn their interest payment with only a fraction of the time. The AI algorithms also are likely to be more accurate and less error-prone, which will result in them saving lots of money in the long run based on the behavior of the human loan agent.

Customers are also going be able to gain. These systems will speed the time needed to go through an application for credit or a loan. What might take a human loan officer several days to go through could take only some minutes time for an algorithm process. The borrower will be informed in a shorter time when they've been accepted for the amount of the loan or the credit card they wish to use.

Supports in the management of fraud and detection

With the many kinds of transactions that happen through financial institutions regularly as well as the various people, loans, and much more they are required to manage so it's not surprising that banks as well as other financial institutions will begin to be worried about fraud, and what it could cost the institution if they don't stop it quickly enough.

Each business, regardless of whether it's an institution of financial origin or not, has to be able to identify ways to minimize the risks of risk. This is particularly true for banks and the various transactions they manage frequently. If you're given an amount of money regardless of whether it's to purchase a house or automobile, or even the credit card it is basically lending you money that belongs to someone else. This is the reason why when you own an account with a savings or checking account or other investment, you're likely to accrue interest.

Another reason that these banks are likely to be extremely vigilant regarding fraud. AI will remain on top of the line in helping companies detect the possibility of fraud as well as any security concerns. AI is able to analyze every spending pattern that were normal and permitted before, as well as using different tools, to identify when there's something that is a little unusual and should be investigated and even put on hold. For instance, if a credit card comes from someone from Nebraska and was utilized there at 10 AM, but the next morning, the transaction took place in Germany It is probable that fraud is occurring which is why the AI tool can identify this.

Perhaps it is because the AI algorithm is taking a look at an account and has a accurate idea of how the user withdraws funds in the first place, what amount, when the time, etc. There comes a point when someone else attempts to withdraw money that is odd for this type of account, or withdraw even though the account holder was already in the same place for the entire day. These are situations can be solved by the AI system of this field can assist with.

Another great benefit you can take advantage of in the field of AI is the fact that it will not encounter any difficulties or concerns in the process of learning. It will be able to flag the red flags for some routine transactions, and often, the system is not in alignment with. Then, a human step in, observes it's legitimate and rectifies that. Instead of being concerned about doing things incorrectly and making errors like humans do it learns from its mistakes and incorporates it into its understanding. In time, these mistakes can aid the algorithm in its ability to come to decisions more sophisticated in determining what qualifies as fraud and what isn't and ensuring that the institution's financial institution can stay clear of this problem at any cost.

Financial advisory services

The other thing we should be looking at the way in which AI can assist with various advice services for financial advisors. In accordance with PWC's PWC report, consumers are advised to consider using more advisors who are robots when they are working with their banks or different financial institutions. With greater demand on banks to cut down on the amount of commissions they charge on investments of individuals It is possible for the robot to perform things that no human would do; make a down payment.

Another area that is beginning to gain momentum in this regard is the bionic advisory. It will be able to bring together human insight with machine-generated calculations to offer alternatives that will be more effective than human insights to give us choices that are more effective than what each component can do by themselves. Collaboration to get this to work, however is going to be crucial.

What we're saying here is that it's not enough for us to simply take an in-depth look at the machine and view it as an add-on or as a system that knows it all for every part. It is necessary to find an appropriate balance between both, and we must be able to think of AI more as an element in the process of making decisions which is equally significant as what humans can contribute in the process. If we can achieve this, then we'll can see the inside about the future for decision-making in the world of finance.

Trading

Also, we can examine the way artificial intelligence will be able to aid in the world of finance to be in the area of trading. Investment firms have relied on data scientists as well as computers to assist them in determining some of the patterns in the future that are exhibited in the markets. As a field of trading and investing, the future is going to rely on the individual investor or machine to forecast the future in a reliable way to make more money from the process.

Machines can be an excellent tool for all of this since they can process a huge volume of data and sift into it within a brief period of time. We can also teach machines how to spot patterns that have occurred with the data over time, and then use that information to come up with good predictions of what patterns might be repeated in the near future.

Although there are some atypicalities that are associated with this instance, like the 2008 financial crisis but it is still feasible to instruct a machine to analyze the information that is provided to identify the triggers that typically occur before the anomalies and then prepare for the future by using the appropriate type of forecasting. What's more important is that, based on the amount of risk an individual is willing to risk, AI is able to use the information to suggest options for the portfolio that are capable of meeting their needs.

Consider the ways this technology will transform the way we invest. Anyone with an increased risk tolerance will be able to trust an AI machine to assist them make the right choices about when to invest or hold or when they should sell their stocks. However anyone who is at the lower side of the risk of investing will get some notifications in the event that it is anticipated that the market is likely to decline, and will then be able to make better choices about whether to stay in the market or change things according to their personal preference and risk-appetite level.

The management of finances

The fifth field we can explore when we are considering making use of AI in the field of finance is aiding people in managing their own financial affairs. managing finances in a world that is constantly connected and materialistic could be an obstacle for all of us. As we look at how it could alter in the future, it's clear the ways in which AI can help assist us in managing our money.

PFM, also known as personal financial management also known as PFM, is expected to be among the latest developments associated with wallets that are accessible online and built on AI. One example is Wallet that was founded at the time of San Francisco, is able to make use of AI to develop a variety of algorithms to make sure that customers are able to make smart choices about their money and how they'll spend it.

The concept behind this wallet will be straightforward. It is able to accumulate the data accessible from your online footprint It can then make a graph out of every purchase you've made. This will give an insight into the way you spend your money as well as the places you've been with regards to shopping and other. This can help you to see how you can reduce your spending and how you could be able to make a difference.

There are people who don't like this due to their concerns about privacy, however this could be a huge aspect to the future development of financial management in a variety of ways. Therefore, it's likely assist you to organize your finances and make it easier for analysis, without having to keep track of your finances through lengthy spreadsheets or writing it on paper. It doesn't matter if you require help monitoring your investments and observing the way trading works, or you're looking to better manage your finances a little better, this is going to be something you should keep an eye out for using the aid of various AI tools.

It is without doubt that AI will definitely play a major role in how the financial industry is changing in the coming years. Due to the speed at which it is taking incremental moves towards making the various financial processes more user-friendly for customers.

Soon it will collaborate with humans to create solutions that are more efficient and quicker.

Bots will evolve slowly as more and more inventions will come from AI. AI sector. Huge investments are being made by firms that see it as a fantastic method to reduce costs in the long run. Additionally, it could help companies save dollars and time when as compared to hiring human workers. When you consider that human error could be diminished through AI instead of employing an employee who is human and it's not surprising that AI is helping to create the future we will see in the world of finance.

Although the use of AI is only beginning to grow and has plenty of speed to catch up to it, there's plenty of potential in this area and a lot of financial institutions are attempting to incorporate AI into their daily tasks. It is likely to happen to come in the near future when AI as well as the technology that go with it will result in small loss (especially when compared to the way humans handle the job) as well as better training and possibly the best customer experiences about.

ARTIFICIAL INTELLIGENT

Many businesses are plagued by monotonous boring, monotonous, and difficult processes, which can delay output and increase the cost of operations. Companies have no other choice than to automatize certain operations to lower production costs.

Through the digitization of repetitive processes businesses can cut expenses associated with paper and work, as well as reduce human errors and errors, which can lead to increased efficiency and better outcomes. In order to reap the advantages described above, companies must select the most appropriate automation tools as otherwise, the efforts they put into it will be in vain.

Automating processes involves using artificial intelligence systems that aid in digitization and deliver the same or better outcomes as humans would.

Artificial Intelligence (AI) is the process whereby an artificial intelligence machine replicates brain functions that are associated with human brains, such as problem-solving and learning thinking and problem solving the representation of knowledge, intelligence social overall intelligence.

AI's main concerns are reasoning and knowledge, planning, learning natural language processing and perception, as well as the ability to manipulate and move objects. The use of statistical methods including computer-aided intelligence (CAI), soft computing and the traditional symbology are just a few examples of methods.

AI utilizes a variety of tools, such as variations of mathematical and search optimization, logic, and economics-based strategies.

WHY SHOULD YOU STUDY ARTIFICIAL INTELLIGENCE?

The study of artificial intelligence enables students to become software engineers that specialize in neural networks quantum artificial intelligence Human-machine interfacings. Students can also become software engineers for companies creating shopping recommendations as well as analysing huge datasets. A degree in artificial Intelligence opens the doors to a profession in the field of hardware engineering, who develops robots to support home use or electronic parking devices. Artificial Intelligence as a subject wasn't even in existence ten years ago, and it is still developing today. AI is believed to be capable of solving some of the world's current and future problems. It is an expanding field with a wealth of opportunities that are likely to be more common in the near future.

WHAT ARE THE FOUR DISTINCT CATEGORIES OF ARTIFICIAL INTELLIGENCE?

As per Arend Hintze who is an Assistant Professor of Integrative Biology as well as computer engineering in Michigan State University, AI can be divided into four categories starting with task-specific intelligence systems that are commonly used currently and moving on towards sentient machines that don't yet exist. The categories are as follows:

Type 1 Reactive machines. Artificial intelligence systems have no memory and are focussed on task. Deep Blue, the IBM program for chess that beat Garry Kasparov in the 1990s is a prime illustration. Deep Blue is capable of finding pieces on the board and making predictions, however it has no memory, and is unable to draw from previous experiences to influence future games.

* Type II: limited memory. AI systems have memory, which allows them to draw from previous experiences to guide future decisions. This is the way a part of the decision-making systems that are used in autonomous vehicles are developed.

"Type 3": Theoretical Mind. The term "theory of mental faculties" can be used to describe the field of psychology. If applied in the context of artificial intelligence this implies that the machine has the social intelligence required to be able to recognize emotions. This type of AI can be capable of deducing human intentions and anticipating behavior that is essential to allow AI systems to be able to function as an integral part team members with human counterparts.

* Type 4: Self-awareness. In this type artificial intelligence systems have an inner self which gives them consciousness. Self-aware machines are aware their present situation. This kind of artificial intelligence doesn't yet exist.

WHAT ARE THE APPLICATIONS OF ARTIFICIAL INTELLIGENCE?

Artificial intelligence isn't something that is a futuristic idea It is in fact, already present and being utilized in a variety of areas. Criminal justice, finance as well as national security, health care transport, as well as intelligent cities can all be covered within this class. There are numerous examples to show that AI is already making significant influence on our world, and substantially enhancing human capabilities. [

One of the main reasons for AI's rising popularity is the huge potential for economic growth it offers. According to a PriceWaterhouseCoopers study, "artificial intelligence technology might boost global GDP by $15.7 trillion, or 14%, by 2030." That includes $7 trillion of advances that are being made in China, $3.7 trillion in North America, $1.8 trillion in Northern Europe, Africa and Oceania are worth $1.2 trillion while the remainder of Asia beyond China has a value of $0.9 trillion. Southern Europe is worth $0.7 trillion while Latin America is worth $0.5 trillion. China is making incredible advances as it has set a target for its national economy to invest $150 billion into AI by the year 2030, and becoming the leading country in this area.

A study carried out by McKinsey Global Institute China discovered that "AI-driven automation can boost efficiency in the Chinese economy which can add 0.8 or 1.4 percent to the annual growth in GDP depending on the pace that it is adopted." While the authors of the report discovered the fact that China is currently far in comparison to in comparison to the United States and the United Kingdom in regards to AI adoption, vast size of China's AI sector offers huge opportunities for pilot testing and further technological advancement.

Artificial Intelligence has impacted many markets. Below are 11 (11) examples.

ARTIFICIAL INTELLIGENCE IN HEALTHCARE.

Artificial intelligence tools aid developers in increasing the complexity of computation in healthcare. Merantix is one example. It is an example of a German company that makes use of deep learning to tackle medical issues. It is utilized in medical imaging with the goal for "detection of lymph nodes within the human body with Computer Tomography (CT) images." According to the creators of the technology, labeling the lymph nodes and identifying any possible growths or lesions that could be dangerous is essential. Humans can accomplish this, but radiologists cost $100 an hour and can only be able to read four images each hour. If they need to read 10,000 photos the procedure will be $250,000. This is costly if done by hand.

What deep learning is able to accomplish in this situation is to teach computers massive data sets to identify the distinction between a normal-looking lymph node and an odd-looking lymph node. After having completed the imaging exercises and enhancing their accuracy in labeling radiological imaging specialists are able to apply their new information to assess patients and pinpoint the degree to which a person is at risk of developing malignant lymph nodes. Since only a few will test positive, the process of determining the healthy node from the healthy one is easy.

Furthermore artificial intelligence has also been used to treat congestive cardiac malfunction, a condition that is a problem that affects 10% of older people and cost in the United States $35 billion each year. Artificial intelligence systems are beneficial as they can 'predict ahead of time potential problems and allocate funds for training, monitoring and proactive steps to ensure that patients stay away from hospitals.'

The biggest bets are focused on improving the outcomes of patients and cost reduction. Companies are using machine learning to provide more precise as well as timely diagnosis than doctors. I.B.M. Watson is among the most popular healthcare technologies. Watson is able to comprehend natural language and can be capable of responding to questions that are posed to Watson. The system analyzes patient information and other publicly accessible sources of data to create the hypothesis, and then provides an confidence scoring scheme. Other AI applications involve the use of chatbots and virtual health assistants to aid health care customers with finding medical information, making appointments, understanding the billing process, as well as performing other administrative tasks. In addition, various artificial intelligence technology is being used to predict, fight and study pandemics like COVID-19.

Healthcare Benefits of Artificial Intelligence

It can be used to detect the genetic code's connections to using surgical robots, and even increasing efficiency of hospitals, AI has proven to help the health industry.

1. Assistance in Clinical Decisions

It is essential for healthcare professionals to take into consideration every crucial piece of information while diagnosing patients. Therefore, sifting through the numerous notes that are not structured found in medical records is vital. If even a single crucial element is missed and the patient's health could be at risk.

Natural Language Processing (NLP) assists doctors in identifying the most pertinent information from the patient's reports.

Artificial intelligence is able to process and store huge amounts of information, which can be used to create knowledge databases as well as the individual evaluation and suggestions of every patient, thus assisting in the development of the clinical decision-making process.

Doctors can count on this technology to aid them in identifying warning signs of danger in notes that are not structured. One fascinating example is IBM's Watson which uses AI to anticipate heart failure.

2. Chatbots can be used to enhance medical care, and to triage

Many people reserve their appointments with their doctors for the smallest health risk or issue, which may often turn into a false alarm or cause for a remedy them.

Artificial intelligence can ensure that primary care is efficient and automated, which allows doctors to take on more demanding and irritated patients.

To cut down on the cost of unnecessary appointments with a doctor Patients can make use of chatbots that are medical, powered by artificial intelligence and come with sophisticated algorithms that provide patients fast responses to their health-related concerns and questions and also provide advice on how to handle any issues that might arise.

Chatbots are accessible round the clock and are able to handle multiple patients at once.

3. Robotic Surgical Procedures

Robotics and artificial intelligence have revolutionized surgical procedures by increasing the capacity and speed while making delicate incisions. Since robots aren't tired, the problem of exhaustion over long and complex procedures is avoided.

AI machines can develop new surgical methods by analysing information from previous procedures. The precision of this device will eliminate the possibility of tremors and unwanted or accidental motions during surgical procedures. Some examples of surgical robots are Vicarious Surgical, which combines VR and AI-enabled robotics to allow surgeons to perform minimally invasive procedures. Also, Heartlander A miniature mobile robot created in Carnegie Mellon's robots division to assist in the heart's therapy.

4. Virtual nursing assistants

Artificial Intelligence technology allows to create virtual nurses capable of carrying out a range of duties, from interfacing with patients, to directing patients to an suitable and efficient healthcare facility. Virtual nurses are available 24/7 and can respond to questions and also assessing and managing patients.

A variety of AI-powered virtual assistant apps already provide more frequent interaction between caregivers and patients in office visits, which can help avoid unnecessary hospital visits. Care Angel, the world's first virtual nurse assistant is also able to conduct health checks with the use of voice and artificial intelligence.

5. Assistance with the diagnosis

AI is able to surpass human doctors in regards to diagnosing, forecasting, and diagnosing ailments more accurately and quickly. In the same way, AI algorithms are accurate and precise at the specialty diagnosis level , and cost-effective in diagnosing diabetic retinal diseases.

PathAI is one example. It is currently developing machine-learning technologies to aid pathologists in making more precise diagnosis. The company's goals currently include reducing diagnostic errors in cancer and establishing methods to provide customized medical treatment.

Buoy Health is an artificial intelligence-powered symptom and cure-checker which diagnoses and treats illness with algorithms. It works like this the chatbot is able to listen to the symptoms of a patient and health issues. The chatbot then guides the patient to the right treatment based on the diagnosis of the chatbot.

6. Reduce stress that comes with EHR usage

While EHRs have been a crucial element in the industry's shift to digitalization, their use has presented a number of challenges by cognitive overloaded, interminable paper work, and burnout.

The EHR developers are now to incorporate AI to design user interfaces that are more intuitive and to automate some of the routine operations that consume a large amount of the time.

Though voice recognition, dictation and voice recording assist during recording medical records but neural process of language (NLP) methods may fail. In addition, AI can assist in handling common requests in the inbox for refills of medications as well as notifications. It can also assist in prioritizing items that require the attention of the doctor, allowing the users to organize their tasks better.

What are the risks to the healthcare industry through Artificial Intelligence?

According to an Brookings Institution paper, various issues related to AI in the healthcare industry need to be addressed. Below are some of the risks outlined in the report by the Institution:

1. Errors and Injuries

The possibility it is possible that an AI system is sometimes in error is among the main concerns associated to AI. For instance, if it recommends the wrong medicine to a patient or make an mistake when identifying an abnormality during an imaging scan, it could cause the patient to be injured or suffering serious health effects.

AI mistakes can be distinct because of at least two reasons. Although human medical experts may make mistakes too however, what makes this crucial is that the root cause or error, within the AI system, can cause damage to many thousands.

2. Data availability

Another threat the threat of AI systems is the fact that they require huge quantities of data from various sources, including information about pharmacies as well as electronic health records as well as insurance claim records.

Since data is scattered and patients often see multiple doctors or transfer insurance companies Data becomes muddled and more difficult to understand which increases the chance of inaccurate information and the cost of collecting data.

3. Privacy concerns

The accumulation of huge datasets as well as the sharing of information between health facilities along with AI developers to facilitate

AI systems can lead patients to think that this might compromise their privacy, leading patients to bring lawsuits.

Another area in which the application of AI systems raises this concern is the possibility of AI to identify patient's private information even if the patient not provided it.

For instance, Parkinson's disease can be detected through an AI system that relies on the trembling sound of an electronic mouse, even though the patient hasn't disclosed the data to anyone else, something that is an invasion of privacy.

4. Inequality and discrimination

Since AI systems learn from the data they use to create their models, they may as well absorb biases that are inherent in the data available. In particular, in the event that data used to create AI is mostly gathered from medical schools and hospitals, the resulting AI systems will not be conscious of, and thus less effective in treating patients who are not frequent visitors to hospitals that are academic.

5. Could cause modifications to the field of work

In the future the usage of AI systems could lead to changes in the medical profession. Particularly in areas like radiology, where the bulk of the work is automated.

This raises the possibility that the usage of AI could lead to an erosion of human understanding and capability as time passes, preventing doctors from identifying AI weaknesses and hindering medical research.

Conclusion

While the technology remains in shadows over specific threats and dangers but it will aid the medical profession through faster treatment and more precise diagnosis and data analytics to detect patterns or genetic data which could lead to a particular disease.

Today, we live in a world where even a few minutes could save lives. Machine learning and AI are able to transform not just the health of individuals but also the lives of every individual.

ARTIFICIAL INTELLIGENCE IN BUSINESS.

In both analysis and customer relation management (CRM) platforms machines learning algorithms are integrated to give information on how customers can be better provided with. Chatbots are integrated into websites to provide customers instant support. Automation of tasks is also a subject of debate between academics and IT specialists.

ARTIFICIAL INTELLIGENCE IN EDUCATION.

Grading could be automated by AI which can free teachers the time of teachers. AI can assess students and modify to their specific needs and allow students to take their time and work at their own pace. AI tutors are able to supplement students' efforts and assist to keep them on the right track. Technology has the potential to transform the ways students study, perhaps displace certain professors.

ARTIFICIAL INTELLIGENCE IN FINANCE.

Within the United States, investments in financial AI increased by more than three times from 2013 to 2014 and reached $12.2 billion. According to industry experts, "loan decisions are increasingly made by algorithms that take into account an array of carefully analyzed information about the borrower instead of only a credit score and an investigation into their background." There are also Robo-counselors who "build individual investment portfolios that are customized which eliminate any need to have stockbrokers or financial advisors." These innovations are designed to eliminate the emotion associated with investing and allow investors to make rational choices within minutes.

An excellent example can be seen on stock exchanges where high-frequency trading that is automated has mostly replaced human decision-making. Customers make buy and sell requests and are immediately matched by computers with no human input. On a tiny scale computers can detect the market's inefficiencies in trading and then execute profitable trades on the basis of the instructions of investors. With certain areas being powered by computers of the present and software, these instruments have dramatically greater storage capacity due to their dependence of "quantum bits" that are able to hold multiple values in every location. This greatly increases the storage capacity and increases processing speed.

Fraud detection is a different application of AI in the field of financial systems. While detecting fraud in large companies can be difficult, AI can discover anomalies and outliers or unusual instances that require further investigation. Managers are able to spot problems early on in the process before they become harmful levels.

AI has a major impact on financial institutions , particularly in personal finance applications like Intuit Mint or TurboTax. These types of apps collect personal data and provide financial tips. Other tools, like I.B.M. Watson are also used to aid in the home buying process. Today artificial intelligence software is responsible for most of Wall Street trading.

CRIMINAL JUSTICE

Artificial Intelligence is used within the justice system for criminals. In the city of Chicago has developed the artificial intelligence driven "Strategic Subject List" that evaluates those who are detained in the hopes of becoming future criminals. It assigns a number to 400,000 individuals based upon their age, crime conduct and victimization, arrest for narcotics background, and gang affiliation. The analysts found that youth is an effective prediction of the likelihood of violent crime. The fact that a victim of a shooting increases the chance of being a perpetrator in the future The gang's involvement in crime does not provide any predictive value and that drug arrests don't significantly predict the future of criminal activities.

Judicial experts say AI programs can reduce human bias when it comes to law enforcement which results in a more fair sentencing system. Caleb Watney, an associate at the R Street Institute, writes:

Questions about predictive risk analysis that are based on empirical research make use of the advantages that machine-learning, automatic reasoning and various other types that use AI According to one machine learning policy simulation, these kinds of software could help to reduce criminality by as much as 24.8 percent, while keeping current jailing rates in place or to decrease the population of jails by as much as 42 percent, while maintaining levels of crime.

However, critics worry that AI algorithms are an "secret system to punish citizens for offenses they haven't committed." Many times, score of risk has been employed to direct massive roundups." The concern is that these tools are unfairly targeting people of color , and haven't been able to assist Chicago in resolving the recent wave of murders.

However however, other countries are speeding up the deployment of this technology. In China for instance companies already have "considerable resources" and have access to facial, voice as well as other biometric data which will assist in the development of their technology." The latest technologies allow the matching of images and voices to other kinds of data as well as the application of AI to combine these data sets to enhance national security and law enforcement. Chinese law enforcement agencies are making an "police cloud" consisting of videos and social media interactions and travel information, online purchases as well as personal information by using their "Sharp Eyes" program. Authorities can make use of this integrated database to keep an eye on potential criminals, terrorists, and lawbreakers. In terms of, China has surpassed the United States as the world's top AI-powered surveillance nation.

ARTIFICIAL INTELLIGENCE IN LAW.

In the legal field the process of discovery -- the process of sifting through documents often overwhelming for human beings. Making use of AI to help automate work-intensive tasks in the legal industry helps to reduce time and improve client service. Legal firms employ machine learning to explain the data and anticipate outcomes Computer vision is utilized to sort and extract information from documents. Natural processing of language is utilized to interpret requests for information.

It is law that is the force which ensures that society has its properness and dignity. There isn't a single sector of industry that has been in a state of non-responsibility to the laws. Every single thing requires legal binding contracts, from selling to purchasing mergers, acquisitions and partnerships.

It is impossible to make growth or advancement without a strong intellectual property law system. Every single one of our activities is under the shadow of the judiciary system that acts as a constant threat to our security.

In this regard, a very robust sector, the legal industry is in no way immune to the technological power which is continuously advancing across its various spheres. Absolutely technological advancements in the field of law have led to the growth of the activities of lawyers.

Because of the growing automatization of legal processes Legal professionals like lawyers and paralegals are required to master areas like communication, word processing and the presentation of data.

Technology for law has affected every aspect that comprise the profession including corporate and law firms as well as courtroom procedure and record management. Artificial intelligence and the advancement of big data make it possible for current software to read legal documents, simplify communications and provide suitable casework for lawyers.

According to McKinsey approximately 23percent of tasks that lawyers do can be automated by using current technologies.

ARTIFICIAL INTELLIGENCE IN MANUFACTURING.

Manufacturing has set the standard in the integration of robots into workflows. For example, industrial machines which were designed to complete a single task and were separated from humans are now becoming cobots: small, multitasking robots that collaborate with humans and assume more responsibility in factories, warehouses, floors, and other workplaces.

ARTIFICIAL INTELLIGENCE IN BANKING.

Banks have successfully utilized chatbots to help customers understand their services and products, and handle transactions that don't require the involvement of humans. Artificial intelligent virtual assistants are employed to simplify and decrease the costs of conforming to the banking regulations. Banks are also using AI to enhance lending decision-making, create credit limits, and find investment opportunities.

ARTIFICIAL INTELLIGENCE IN TRANSPORTATION.

Transportation is a field in which AI as well as machine-learning are driving significant advances. Between August 2014 to June 2017 the Brookings Institute's Cameron Kerry and Jack Karsten found that more than 80 billion dollars were invested in autonomous car technology. These investments include autonomous driving technologies as well as the technology that drives the industry.

Autonomous vehicles - cars, trucks, buses, as well as drone delivery systems rely on the latest technology. These include autonomous vehicle braking and steering and lane-changing systems, sensors and cameras for accidents prevention, real-time data analysis through artificial intelligence and the use of advanced computing technology and deep learning systems that can adapt to changes in the environment using precise maps.

Navigating and collision avoidance need light detection and range systems (LIDARs) as well as artificial intelligence (AI). LIDAR systems incorporate the radar and light beams. They are mounted on the top of vehicles and are able to measure the distance and speed of nearby objects by capturing images of them in the 360-degree space created by light beams and radar. With sensors on the sides, the front and back side of the automobile, the devices provide information that aids moving vehicles and trucks in keeping their lane clear, to avoid collisions with other vehicles, applying brakes, and change direction as needed and in a timely manner to prevent accidents.

Because of the large quantity of data gathered by these sensors and cameras and the need to process it immediately so as to avoid colliding with a vehicle in the next lane autonomous cars require processing that is high performance, sophisticated algorithms, and deep-learning systems to be able to adapt to the changing situations. The key to success lies with the program, and not in the vehicle or truck itself. Modern software allows vehicles to take the lessons of road users, and to modify their systems of guiding to changes in conditions of the weather, driving and road condition.

Businesses that offer ride-sharing are keenly attracted by autonomous cars. They are looking for the potential to enhance the quality of service to customers and increase productivity. Each of the major ride-sharing companies is studying driverless vehicles. The growing popularity of car-sharing taxi and ride-sharing services, such like Uber as well as Lyft across the United States, Daimler's Mytaxi and Hailo in the United Kingdom, and Didi Chuxing in China, illustrate this new way of traveling. Uber has struck a deal with Volvo to purchase 24,000 autonomous vehicles to use for their ride-sharing services.

But the ride-sharing service was hit hard in the month of March in 2018, when one driverless vehicle crashed and killed an innocent pedestrian in Arizona. Uber and a number of automakers stopped testing right away and launched an investigation into what went wrong and the cause of the crash. Consumers and the industry alike must be assured of the fact that AI is secure and is able to live the promises it makes. If there aren't convincing proofs this disaster could stop AI advancements in the transport sector.

In addition to its crucial function in the operation of autonomous vehicles, AI technologies are employed in transport to regulate travel, anticipate delays at airlines and enhance the security and efficiency of ocean transportation.

SECURITY.

Artificial intelligence plays an important part in the national defense. It is the reason that American military is currently deploying AI by using Project Maven to "sift through vast amounts of information and footage collected by surveillance, and later inform human experts of changes or strange or suspicious behaviour." The aim of new technologies in this area according to the Deputy Secretary of Defense Patrick Shanahan, is to "meet the demands of our soldiers while increasing the speed and speed for technology-related development as well as procurement."

Big data analytics paired to AI can significantly affect intelligence analysis since huge quantities of data will be processed in near-real-time and, if not in real-time, providing the commander and his staff to attain previously unimaginable level of intelligence analyses and efficiency. Command and control can be affected when commanders from human control outsource routine or, in some instances critical decisions in order to AI platforms, drastically decreasing the time needed to make a decision and the subsequent actions. In addition combat is a time-sensitive activity with the side which can take the fastest decision and implement the most quickly is likely to win. In fact, when paired with AI-assisted command and controlling system, AI-assisted command and control systems can speed up decisions and support to levels that are far greater than conventional combat. The process is very rapid, particularly when it is combined with automated decision-making to deploy artificially-intelligent autonomous weapons systems that can produce fatal results that a new term, hyperwar is being used to describe the speed at which war is conducted.

While the legal and ethical debate continues on what happens if there is any chance that United States will ever engage in war using AI autonomous lethal systems, Russians and Chinese are not as entangled in the discussion, and we must be prepared to protect ourselves against systems operating at hyper-war speed. The Western world's battle to figure out the best way to place "people who are in the loop" in a hyper-war situation will ultimately determine the ability of the West be competitive in the emerging kind of war.

In the same way that AI is expected to have a major impact on the speed of warfare and effectiveness, the rise of zero-day cyber threats as well as polymorphic malware will make even the most advanced cyber defenses based on signatures to the examination. This requires a significant improvement to the current cybersecurity defenses. Systems that are vulnerable are evolving in a rapid manner and require a multi-layered approach to cybersecurity that utilizes cloud-based intelligent AI platforms. This approach improves the security capabilities of the community towards an "thinking" defense capability that can defend networks by continuously educating themselves on threats that are known. This involves the analysis at the DNA level of undiscovered codes and the capability to block potentially dangerous inbound codes based on the string content of a file. This is the way that certain crucial technologies in United States could thwart the destructive "WannaCry" as well as the deadly "Petya" virus.

Prepare for hyperwar and protect important cyber networks . This should be an important priority in the same way that China, Russia, North Korea as well as other nations invest significant funds into artificial intelligence. The Chinese State Council announced a plan in 2017 to "create an industry at home that will be worth around 150 billion dollars" in 2030. To illustrate the potential, Chinese Baidu search engine Baidu was the first to use facial recognition to identify missing people. In addition, cities like Shenzhen are providing AI laboratories with up to $1 million. They believe that artificial intelligence can improve security, reduce terrorism and help improve speech recognition systems. Since there are many AI algorithms are multi-purpose research, the findings from one area of society could easily be adapted for the security industry.

The buzzwords of today to differentiate security solutions comprise machine learning and artificial intelligence. In addition, these terms refer to technologies that are feasible. Machine learning is employed by companies working in security information and management (SIEM) software and other domains to spot suspicious activities and anomalies that are that could indicate a threat. Through analyzing data and applying algorithms to identify similarities with previously detected dangerous software, AI can send alerts to new and emerging threats faster than human workers or previous technological variations. The advancement in technology is crucial to assist businesses in protecting against cyber-attacks.

INTELLIGENT CITIES

Metropolitan governments are using artificial intelligence to improve the quality of urban services. For example, Kevin Desouza, Rashmi

Krishnamurthy as well as Gregory Dawson report that the Cincinnati Fire Department uses data analytics to improve the response time to medical emergencies. The new analytics system will advise the dispatcher of the most appropriate response for an emergency medical call, determining whether the patient is able to be treated on the spot or must be transported to a hospital taking into consideration a variety of factors such as the nature of the call and the location, weather, and similar calls.

Since the city receives 80,000 inquiries each the year Cincinnati authorities are making use of this technology to prioritize their responses and determine the most effective option in the situation in the event of an emergency. They consider AI as a means to handle massive amounts of data and develop effective responses to the demands of the public. Instead of addressing issues with service on the fly, they attempt to take a proactive approach to urban services.

Cincinnati isn't the only city in this area. Many cities are adopting smart city apps that make use of artificial intelligence to improve the quality of services environmental planning resource management efficiency, energy efficiency, and the prevention of crime in addition to other functions. Fast Company assessed American cities to determine their smart cities index and determined New York City, San Francisco, Boston, Seattle as well as Washington, DC as the most popular users. Seattle. For instance, it has adopted sustainability as a goal and uses artificial intelligence to manage its use of energy and resources. Boston has created an"mobile "City Hall" to ensure that the most disadvantaged communities can access essential municipal services. In addition, it has installed "cameras along with inductive loops to assist with traffic management and also sound sensors that detect shootings." The buildings of San Francisco have been recognized for their compliance with LEED sustainability standards.

By implementing these strategies, and others With these and other strategies, metropolitan cities are ahead of other regions of the nation in regards to AI deployment. In fact more than 66 percent of American cities, as per the National League of Cities survey have invested in intelligent city technology. The report highlights a variety of notable applications, such as "smart meters for utility services smart traffic signals, intelligent traffic sensors E-government apps Wi-Fi kiosks and RFID sensors incorporated into pavement."

ARTIFICIAL INTELLIGENCE'S HISTORY AND EVOLUTION

AI is omnipresent in our modern times in everything from education to healthcare manufacturing laws, finance, and even in politics. But, to think that AI is a distinctly modern technology would be an error.

Incredibly, the idea of AI was gradually expanding in the minds of our forefathers over an extended time as classical philosophers understood human thinking as a symbol system. Robots were a beloved myth in the time of the early Greeks and the first references to automata go way back Greece in the Middle Ages and China.

Did you be aware it was John McCarthy coined the term artificial intelligence in 1956, during an Dartmouth College conference?

From the 1950s to the present the various researchers, programmers, and theorists helped to develop the modern idea of AI through consistently creating new ideas in the field, changing it from a fantasy of our imagination into a real possibility.

This chapter will examine a variety of notable examples that have influenced the evolution and development in the field of Artificial Intelligence.

Artificial Intelligence Timeline

It is possible to believe that the story of AI began in 1956 , when John McCarthy coined the word but the tale of AI's growth began before that.

AI in 250 BC

The truth is that it all started in the year 250 BC at the time that Ctesibius who was a famous Greek scientist and inventor created the world's first self-regulating artificial system. The system, named"clepsydra" or "water thieves," was designed to ensure that the reservoir that was used to make water clocks (clocks made to display the time of day) was always filled.

AI from 380 BC - the late 1600s

In this time, a variety of mathematicians, theologians, and philosophers wrote books on the mechanical process and number systems. They also developed the notion that mechanical "human" thinking within non-human objects. For instance, Ramon Llull, a Catalan poet, theologian, and poet created Ars generalis ultima (The Ultimate General Art) which honed his understanding of mechanical methods based on paper to produce new knowledge by conceptual combination.

Ai in the Early 1700s

In the novel of Jonathan Swift "Gulliver's Travels" an invention known as the engine was featured in the novel, which was one of the first references to the latest technology that was, in this case, computers. The main goal of the project was to increase knowledge and improve mechanical operation to the point that even the weakest person could appear skilled thanks to the expertise and guidance of a mind that was not human and simulated artificial intelligence.

AI in 1872

In 1872, writer Samuel Butler released his novel "Erewhon," in which Butler speculated that machines could someday be able to recognize.

AI from 1900 to 1950

When the 1900s started there was a drastic change in the speed of AI advances took place. The story of AI accidents in the early 20th century can be fascinating.

1919: Karel Capek (born 1921), an Czech playwright, releases the science fiction novel "Rossum's Universal Robots." The play was a celebration of manufactured artificial humans that he describes as robots. This is the first version of the phrase that has been known. After this, a lot of people were drawn to the idea of a robot and began to incorporate it into their artwork and research.

In 1927, the sci-fi film Metropolis was released in 1927 (directed by Fritz Lang). The film is famous for its depiction on screen of a robotic figure, which influenced the next generation of famous non-human creatures.

1929 Makoto Nishimura was a Japanese biologist, professor, created Gakutensoku the first robot. The term means "learning from natural laws," meaning that the artificially intelligent robot learned from the environment and from humans.

1939 The inventor and physicist John Vincent Atanasoff and Clifford Berry who was the graduate assistant created the Atanasoff

Berry Computer (ABC) which was a digital computer that could be programmed located at Iowa State University. The computer weighed more than 700 pounds and could concurrently solving 29 equations in linear form.

1949 Edmund Berkeley, a mathematician and actuary, published in 1949 the work "Giant Brains": or Machines That Think." The book discussed how computers are becoming more adept in managing large quantities of data. In addition, the book compares the capability that machines have to process information with the human brain and concluded that machines are able to think.

AI in the 1950s

The 1950s witnessed a rise in research-based findings on artificial intelligence. This resulted in numerous breakthroughs in the field.

1950: Claude Shannon, dubbed the "father of information theory,"" published "Programming the Computer for Playing

Chess,"" the first piece of writing that describes the development of an algorithm for computers capable of playing Chess.

1950. Alan Turing's paper on "Computing Machines and Intelligence" was one of the pivotal moments in AI the history of AI in the year 1950. The document discussed the idea of "The Imitation Game," later called "The Turing Test" an idea that evaluated the capability of machines to behave like humans. The Turing Test evolved into an essential component of artificial intelligence.

1952. Arthur Samuel invented a program to play checkers. The first machine that could be able of competing with human players in playing Checkers.

1955: Allen Newell (researcher), Herbert Simon (economist) as well as Cliff Shaw (programmer) created the first computer program for artificial intelligence, Logic Theorist, which established 38 of the initial 52 theorems found in Whitehead as well as Russell's Principia Mathematica.

1955. John McCarthy, an American computer scientist, along with his colleagues proposed a conference focused on "artificial intelligence."

1956: After this workshop John McCarthy coined the artificial intelligence term during an Dartmouth College conference.

1958 McCarthy was at MIT developed Lisp which is a top-level AI programing language for research that is being used today in the modern world.

1959 Arthur Samuel coined the phrase "machine learning" while creating a computer for chess which could compete against human gamers. 1958:

AI in the 1960s

The decade of the 1990s saw massive growth in creation of robots and automatons, as well as new language programming, as well as research and films that show artificially intelligent individuals.

1961: In this year's edition George Devol's 1950s industrial robot Unimate was the first robot to operate on the General Motors assembly line in New Jersey. The task of the robot was to perform tasks which could be risky for humans to carry out including transporting die castings from the production line.

1961: James Slagle (the computer scientist and professor) developed his SAINT (Symbolic Automatic Integrator). It is a software that solves the problems of symbolic integration that were encountered in the first year of calculus.

1964 Daniel G. Bobrow Daniel G. Bobrow, an American computer scientist developed the first artificial intelligence program STUDENT written in Lisp. The program was developed to aid students in reading and solving algebraic word problems.

1966 Ironically, the very first chatbot Eliza the computer-generated program that processed natural language created by Prof. Joseph

Weizenbaum at MIT, was originally created as an idiot and not as connecting. However, many of them developed emotional attachments to the program. While Eliza could communicate with humans using text messages and was incapable of learning from humans, her interactions was able to help in removing the barriers to communication between robots and humans.

1966: That was the year that the idea to build the first robot that could move around, "Shakey," began. From 1966 until 1972, this project was in operation. The project was seen as a method of connecting several branches of artificial Intelligence using computer vision and navigation. The robot is in the Computer History Museum.

1968 was the year of the debut of the cult science fiction film 2001: A Space Odyssey. The film, which was directed by Stanley Kubrick, starred HAL (Heuristically programmed Algorithmic computer) which was an artificial intelligence powered by nuclear power on the Discovery One spaceship. The onboard supercomputer was reminiscent of contemporary voice assistants like Siri as well as Alexa.

"That might be among the greatest achievements of the film that it introduced AI to the forefront of public awareness before the release on the initial AI robotics system, Shakey, in 1969,"

-- Murphy, Texas A&M University professor

Terry Winograd, a Stanford University computer science professor, developed SHRDLU the first natural language computer program, in the year 1968.

AI in the 1970s

The decade of the 1990s saw major breakthroughs in the field of automation and robotics. However it was also marked by a number of obstacles like the government's decision to cut the amount of money that was allocated to research into artificial intelligence.

1970 Waseda University in Japan created the first humanoid robot named WABOT-1. It was equipped with features like moving limbs that could be moved and the ability to communicate and see.

1973: The progress of AI has slowed to a crawl as mathematical scientist James Lighthill informed the British Science Council of the status of AI research, stating that none of the results were yet having the expected impact. This led to the British government dramatically cut its funding to AI research.

1977: 1977 marked the start in the history of Star Wars' famous legacy. The film, written and directed by George Lucas, stars C-3PO Humanoid Robot, designed to function as a protocol robot, and predicted to be proficient in seven million different types of communication. It also features R2-D2, a tiny Astromech droid, which is incapable of hearing human speech and communicating via electronic beeps.

1979 1979 Stanford Cart, a remote-controlled TV-equipped mobile robot created in 1961, can successfully cross an entire room filled with seats within 5 hours or less, making it among the first demonstrations of autonomous vehicles.

Artificial Intelligence in the 1980s

AI Winter (a time of lower interest and investments in the area of AI) remains a shadow over the course of this decade. It was then revived as the British government reaffirmed funding to be competitive with Japanese efforts.

1980: Waseda University developed WABOT - 2 that allowed the humanoid to talk with humans as well as read musical charts and play music through an organ that was electronic.

1981: In the name of the Japanese Ministry of International Trade and Industry Around $850 million was earmarked for the Fifth Generation Computer Project. The goal of the project is to create computers that are capable of performing interactions that translate languages, interpret images, and analyzing data the like humans do.

1984 Steve Barron's feature film Electric Dreams is released, an intriguing story of the love triangle of three people: a woman, a male as well as a computer dubbed "Edgar."

1984. Roger Schank and Marvin Minsky warned of the upcoming AI Winter at an Association for the Advancement of Artificial Intelligence (AAAI) meeting. In the following three years, the warning proved accurate.

1986. The year when the first car in the world that was autonomous was a Mercedes-Benz van that was equipped with cameras and sensors and developed by Ernst Dickmanns could reach speeds of as high as 55 miles an hour on empty roads.